# MEDICINAL BIOPROSPECTING: POLICY OPTIONS FOR ACCESS AND BENEFIT-SHARING

# MEDICINAL BIOPROSPECTING: POLICY OPTIONS FOR ACCESS AND BENEFIT-SHARING

APHRODITE SMAGADI

British Institute of International and Comparative Law

Published and Distributed by
British Institute of International and Comparative Law
Charles Clore House, 17 Russell Square, London WC1B 5JP



*British Library Cataloguing in Publication Data*
A Catalogue record of this book is available from the British Library

ISBN 978–1–905221–36–3



Typeset by Cambrian Typesetters
Camberley, Surrey
Printed in Great Britain by the MPG Books Group
Bodmin and King's Lynn

# Contents

# *Tables*

# *Abbreviations and Acronyms*

| | |
|---|---|
| ABS | Access and Benefit-Sharing |
| ABS-MT | Access and Benefit-Sharing Management Tool |
| AG | Aktiengesellschaft |
| AIDS | Acquired Immune Deficiency Syndrome |
| ANDES | Asociación para la Natura y el Desarrollo Sostenible (Association for Nature and Sustainable Development) |
| ARA | Academic Research Agreement |
| ASEAN | Association of Southeast Asian Nations |
| ASP | American Society of Pharmacognosy |
| ASOMPS | Asian Symposium on Medicinal Plants, Spices and other Natural Products |
| BCA | Biological Collecting Agreement |
| BCCM | Belgian Co-ordinated Collections of Micro-organisms |
| BGVS | Bedrijf Geneesmiddelen Voorziening Suriname |
| BMS | Bristol-Myers Squibb |
| BU | Bioprospecting Undertaking |
| CAH | Consejo Aguaruna-Huambisa |
| CBD | Convention on Biological Diversity |
| CGIAR | Consultative Group on International Agricultural Research |
| CGRFA | Commission on Genetic Resources for Food and Agriculture |
| CI | Conservation International |
| CIP | Centro Internacional di Patata (International Potato Centre) |
| CITES | Convention on International Trade in Endangered Species of Wild Fauna and Flora |
| CoC | Code of Conduct |
| COP | Conference of the Parties |
| CPR | Common Pool Resources |
| CRA | Commercial Research Agreement |
| CRADA | Cooperative Research and Development Agreement |
| CSIR | South African Council for Scientific and Industrial Research |
| DA | Department of Agriculture |
| DBFs | Dedicated Biotechnology Firms |
| DENR | Department of Environment and Natural Resources |
| DNA | Deoxyribonucleic acid |
| ed. | Editor |
| eds. | Editors |
| EMAS | The EU Eco-Management and Audit Scheme |
| EO | Executive Order |

| | |
|---|---|
| EC | European Communities |
| EIA | Environmental Impact Assessment |
| EPO | European Patent Office |
| etc-Group | Action group on erosion, technology and concentration |
| FAO | Food and Agriculture Organization |
| PhRMA | Pharmaceutical Research and Manufacturers of America |
| FPIC | Free and Prior Informed Consent |
| GA | General Assembly |
| GATT | General Agreement on Tariffs and Trade |
| GBS | Global Biocollection Society |
| GSK | GlaxoSmithKline |
| GMOs | Genetically Modified Organisms |
| GTZ | Deutsche Gesellschaft für Technische Zusammenarbeit (German Corporation for Technical Cooperation) |
| HIV | Human Immunodeficiency Virus |
| HRC | Human Rights Committee |
| IAEA | International Atomic Energy Agency |
| ICBG | International Conservation Biodiversity Group |
| ImSchG | Immissionsschutzgesetz |
| IAC | Inter-Agency Committee |
| IARC | International Agricultural Research Center |
| ICC | Indigenous Cultural Community |
| ICCPR | International Covenant on Civil and Political Rights |
| ICESCR | International Covenant on Economic, Social and Cultural Rights |
| ICJ | International Court of Justice |
| IGC | WIPO Intergovernmental Committee on Intellectual Property, Genetic Resources and Folklore |
| IISD | International Institute for Sustainable Development |
| ILO | International Labour Organisation |
| INBio | National Institute of Biodiversity |
| IND | Investigational New Drug |
| IP | Intellectual Property |
| IPs | Indigenous Peoples |
| IPRs | Intellectual Property Rights |
| IPRA | Indigenous Peoples' Rights Act |
| IRR | Implementing Rules and Regulations |
| IRRI | International Rice Research Institute |
| ISE | International Society of Ethnopharmacology |
| ITPGRFA | IT International Treaty for Plant Genetic Resources for Food and Agriculture |
| ITQs | Individual Transferable Quotas |
| IU | International Undertaking |

| | |
|---|---|
| IUCN | International Union for the Conservation of Nature and Natural Resources (now The World Conservation Union) |
| IUCN-ELC | Environmental Law Center of the World Conservation Union |
| JUNAC | Junta del Acuerdo de Cartagena |
| LOC | Letter of Collection |
| LOI | Letter of Intent |
| MEA | Multilateral Environmental Agreement |
| MATs | Mutually Agreed Terms |
| MOSAICC | Micro-Organisms Sustainable Use and Access Regulation International Code of Conduct |
| MoU | Memorandum of Understanding |
| MTAs | Material Transfer Agreements |
| NCEs | New Chemical Entities |
| NCI | National Cancer Institute |
| NCIP | National Commission on Indigenous Peoples |
| NGO | Non-Governmental Organization |
| NIH | National Institute of Health |
| NSF | National Science Foundation |
| OECD | Organisation for Economic Co-operation and Development |
| OHCHR | Office of the United Nations High Commissioner for Human Rights |
| OPEC | Organization of Petroleum Exporting Countries |
| OTC | Over-The-Counter |
| para. | Paragraph |
| paras | Paragraphs |
| PAWB | Protected Areas Wildlife Bureau |
| PIC | Prior Informed Consent |
| PLC | Public Limited Company |
| R&D | Research and Development |
| rDNA | Recombinant Deoxyribonucleic acid |
| RAFI | Rural Advancement Foundation International |
| Res. | Resolution |
| RRA | Relative Risk-Aversion |
| PCSD | Palawan Council for Sustainable Development |
| PBRs | Plant Breeders' Rights |
| SBIR | Small Business Innovative Research |
| SEB | Society for Economic Botany |
| SMEs | Small and Medium-sized Enterprises |
| SKB | SmithKline Beecham |
| SMP | Sarawak Medichem Pharmaceuticals |
| SPDA | Sociedad Peruana de Derecho Ambiental (Peruvian Society for Environmental Law) |

| | |
|---|---|
| TK | Traditional Knowledge |
| TRIPS | Trade-Related Aspects of Intellectual Property Rights |
| UDHR | Universal Declaration of Human Rights |
| UIC | University of Illinois at Chicago |
| UK | United Kingdom |
| UN | United Nations |
| UNCTAD | UN Conference on Trade and Development |
| UNCLOS | United Nations Convention on the Law of the Sea |
| UNDP | United Nations Development Programme |
| UNEP | United Nations Environment Programme |
| UNESCO | United Nations Educational, Scientific and Cultural Organization |
| UNHCHR | United Nations High Commissioner for Human Rights |
| UNU | United Nations University |
| UPCH | Universidad Peruana Cayetano Heredia |
| UPOV | L'Union internationale pour la Protection des Obtentions Végétales |
| US | United States |
| USA | United States of America |
| USPTO | United States Patent and Trademark Office |
| WGIP | Working Group on Indigenous Populations |
| WHO | World Health Organization |
| WIPO | World Intellectual Property Organization |
| WSSD | World Summit of Sustainable Development |
| WTO | World Trade Organization |
| WU | Washington University |

# CHAPTER 1

# *The Issue and the Approach of the Study*

## I. THE PROBLEM

At present, a lot of marketed medicines belong to the wide range of products deriving from biodiversity, such as cosmetics, chemicals, and crops. Natural organisms' genetic codes contain the 'recipes' for chemical compounds of potential value for the drug industry, which includes the botanical and pharmaceutical industries. Commercialization of these substances is possible by acquiring a breeding stock of the organism that produces the desired compound, transplanting genes or using the naturally found model for synthesis in the laboratory of the same or related compounds.[1] The botanical industry is examining and using the naturally occurring compounds in unmodified form, whereas the pharmaceutical industry is more inclined to develop 'leads', that is, developing promising molecules of a naturally detected compound that must be modified to increase efficacy or reduce side effects.

Biodiversity prospecting (bioprospecting) is the systematic search for potentially valuable commercial applications for biochemical or genetic resources. Very often, the production of a medicine by private companies is based on genetic resources from plants, animals, microbial sources, marine organisms and insects lying within the regions of rich biota, in countries which usually happen to be developing countries.

Bioprospecting is cited as one of the activities in the use of genetic resources, the other three being customary use and development, public scientific research, and commercial research and development.[2] In reality this categorization, represented in Table 1 on page 2, mirrors the historical movement in the use of genetic resources.

These activities reflect the flow of genetic resources, where their use started with customary use and developments in different parts of the world, and then attracted the attention of the scientific community and researchers. After reassessment of their value, bioprospecting and commercial activities became more and more frequent. According to Table 1 below,

[1] WV Reid et al, *Biodiversity Prospecting* (World Resources Institute, Washington DC, 1993).

[2] M Byström, P Einarsson, GA Nycander, *Fair and Equitable: Sharing the benefits from use of genetic resources and traditional knowledge* (Swedish Scientific Council on Biological Diversity, Stockholm, 1999) 17.

*Table 1: Benefit types and benefit allocation arrangements*[3]

| Category | Customary use and development | Public scientific research | Bioprospecting | Commercial research and development |
|---|---|---|---|---|
| **Benefits** | Knowledge, Use values (food, shelter, health care...), Existence and Option values | Knowledge New use values Academic prestige | Knowledge about commercial potentials (exchange value) | Capital returns (exchange value) New use values (new products) |
| **Benefit allocation** | Formal and informal rules vary for the benefits allocation. Often free within and between communities; ongoing efforts to protect indigenous knowledge | Ethical codes/Public domain (eg scientific publication) | Contracts National access legislation | Intellectual property rights |

bioprospecting lies between public scientific research and commercial research and development (R&D), and it almost always leads to the latter. More specifically, its objective is to obtain knowledge about commercial potentials through the means of contracts and national legislation. It depends on the outcomes of the bioprospecting activities whether commercial activity will be undertaken in order to create and market new products which will bring capital returns. In fact, it is difficult to clearly demarcate between public scientific research, bioprospecting, and commercial R&D, as activities among these three categories are hybridized.

Bioprospecting and commercial activities are rapidly taking over customary and public research use, and from this reality stemmed the access and benefit-sharing debate. There is a problem related to the allocation of the monetary benefits that derive from the sale of these products. Until very recently, intellectual property laws have been the main distribution mechanism for monetary benefits,[4] neglecting the contribution of source countries in matter and knowledge, while non-monetary benefits were not mentioned at all.

[3] ibid 18. [4] See below Chapter 2, Section II.B.

Hence, market forces could not satisfactorily ensure that benefits deriving from commercial use of the results of bioprospecting reached all stakeholders, and particularly those who bear the costs of biodiversity conservation. The market distribution of the benefits of bioprospecting failed to address serious ethical considerations, and led to a loud outcry by the 1980s from developing countries and civil society actors against the so-called 'biopiracy' phenomenon.[5] The term emerged to describe the ways corporations from the developed world claim ownership of, free ride on, or otherwise take advantage of, the genetic resources and traditional knowledge and technologies of developing countries. It was initially invented by the non-governmental organization RAFI, now etc-Group, which started a campaign against the 'illegal' bioprospecting practices in the 1990s. Despite its broad use by NGOs and developing countries, 'biopiracy' remains a rather controversial term.[6]

Patent cases, such as the Turmeric,[7] the Neem,[8] and the Ayahuasca,[9] have received attention in the international press and raised public opinion against the 'bad North and industry' and in favour of the 'good and poor South', whereas various plant resources, like quinine, rubber and cocoa, were 'smuggled' in the past.

The 'biopiracy' debate is linked to the current structure of the domestic

[5] See, eg B Burrows (ed) *Out of Africa: Mysteries of Access and Benefit-Sharing – A Report by Jay Mc Gown* (The Edmonds Institute, Edmonds/USA in cooperation with the African Centre for Biosafety, Richmond/South Africa, 2006), published on the Internet at: http://www.edmonds-institute.org/outofafrica.pdf (08/09/2008).

[6] On the ambiguous meaning of the term, see G Dutfield, *Intellectual Property, Biogenetic Resources and Traditional Knowledge* (Earthscan, London, 2004) 52 and G Dutfield, 'What is Biopiracy?' (paper for the International Expert Workshop on Access to Genetic Resources and Benefit-Sharing, Identification of Outstanding ABS Issues: Access to GR and IPR, Cuernavaca, Mexico, 24–27 October 2004).

[7] Turmeric is used in India for cooking, dying, cosmetics, and medicinal purposes. In 1995, a US patent was granted to two Indian national researchers at the University of Mississippi Medical Center for use of turmeric in wound healing. The patent was successfully cancelled from the US PTO in 1998.

[8] The Neem tree is a quickly growing tree, whose bark, leaves and seeds have various medicinal properties and can be used as a pesticide. Hindus and Muslims consider it a 'blessed tree'; eg twigs are used by Indian peasants as antiseptic toothbrushes and oils as natural insecticide, a contraceptive and in soap. WR Grace, a US-based chemical company worked on a natural insect repellant essence contained in the tree and held US and European patents on the extracting process, which is based on the process previously employed by Indian farmers, without rewarding the local farmers. A number of environmental activists and international environmental groups filed petitions with the US PTO and the EPO seeking to invalidate the patent on the ground that its novelty exists 'in the context of the ignorance of the West'; the patent was finally revoked by the EPO in 2005. See also P Schuler in JM Finger and P Schuler (eds) *Poor People's Knowledge—Promoting Intellectual Property in Developing Countries* (OUP/The World Bank, Washington DC, 2004) 161; CR McManis, 'The Interface between International Intellectual Property and Environmental Protection: Biodiversity and biotechnology' (1998) 76 Washington University Law Quarterly 255.

[9] LM Fecteau, 'The Ayahuasca Patent Revocation: Raising Questions about US Patent Policy' (2001) 21 Boston Third World Law Journal 69.

and international intellectual property regimes. Although the patent system does not allow a naturally occurring substance to be patented ('product of nature' exception), where a substance, 'previously unknown in its purified and isolated form', is refined so that the product can be distinguished in kind, and where it also demonstrates 'unexpected properties', the refined substance is patentable.[10] Moreover, the US patent legislation, defines 'prior art' as an 'invention not known or used in *this country*' (emphasis added), or patented or at least published in a foreign country.[11] Due to the fact that international patent standards in effect build upon patent standards of developed countries, a campaign against the globalized intellectual property system began, which in the case of biogenetic resources and traditional knowledge gave rise to the conflict[12] between the WTO TRIPS Agreement and the Convention on Biological Diversity (CBD).[13]

The access and benefit-sharing provisions of the CBD came to adjust these inequalities and to spell out that 'weak' stakeholders should democratically participate in bioprospecting activities and be compensated for any contribution. Its objectives include the conservation, sustainable use, and fair and equitable sharing of the benefits arising from the utilization of genetic resources. To the end of 'fair and equitable' benefit-sharing, the CBD further requires previous information and prior informed consent of the owner/manager of the resources and an appropriate access and benefit-haring arrangement as mediated through contractual agreements between the owners/managers of genetic resources, on the one hand, and the parties interested in accessing those genetic resources, or studying them, and further exploiting their potential commercial use, on the other.

Efforts to regulate the access to and use of genetic materials are stepping up, at least for those genetic resources within national jurisdiction,[14] with

[10] EJ Asebey and JD Kempenaar, 'Biodiversity Prospecting: Fulfilling the Mandate of the Biodiversity Convention' (1995) 28 Vanderbilt Journal of Transnational Law 711.

[11] See US Act, Title 35, Patents, Part II, Chapter 10, §102 Conditions for patentability; novelty and loss of right to patent.

[12] See below Chapter 4 Section V.

[13] Convention on Biological Diversity, adopted on 5 June 1992 at the United Nations Conference on Environment and Development (UNCED) in Rio de Janeiro, Brazil, 31 Int'l Legal Materials 822 (1992), Articles 1 and 15.

[14] There is a highly uncertain regulatory environment with respect to resources that lie outside the scope of national territories and thus national jurisdiction, such as genetic resources in high seas and deep sea bed or genetic resources in Antarctica and the Arctic. The UNCLOS sets a regime governing mineral resources in the Area under UNCLOS, but further research is needed on whether this regime is adaptable and applicable in genetic resources; see D Leary, *Bioprospecting in the Arctic* (United Nations University/Institute of Advanced Studies, Tokyo, 2008); S Arico and S Salpin, *Bioprospecting of Genetic Resources in the Deep Seabed: Scientific, Legal and Policy Aspects* (United Nations University/Institute of Advanced Studies, Tokyo, 2005); D Lohan and S Johnston, *Bioprospecting in Antarctica* (United Nations University/Institute of Advanced Studies, Tokyo, 2005); and D Lohan and S Johnston, *The International Regime for Bioprospecting—Existing Policies and Emerging Issues for Antarctica* (United Nations University/Institute of Advanced Studies, Tokyo, 2003).

the CBD Bonn Guidelines for genetic resources agreed by the Sixth Conference of the Parties (COP6) to the CBD in 2002 and the Food and Agriculture Organization (FAO) International Treaty for Plant Genetic Resources for Food and Agriculture (ITPGRFA), into force since June 29, 2004, and also with ongoing activities for an international legal framework on animal genetic resources.[15]

The ITPGRFA has the same goals as the CBD, but a completely different approach. The FAO's work commenced much earlier that the CBD. It began in 1983 with the International Undertaking. This is a legally non-binding agreement aiming at the preservation of genetic resources for food and agriculture. The ITPGRFA came to replace the International Undertaking, aligning its goals to the CBD objectives and also has the ambition to put in place a multilateral system to meet these objectives: Material Transfer Agreements (MTAs) are the main tool of transaction for plant genetic resources for food and agriculture in this multilateral framework; however, the users' counterpart is not the providing country, but a central institution responsible for the redistribution of the benefits afterwards, thus opposing the bilateral system introduced by the CBD and the Bonn Guidelines governing all genetic resources within sovereign boundaries.

The CBD proclaims sovereign rights of the States over their biogenetic resources, replacing the previous Common Heritage of Mankind regime and suggests the framework principles and conditions of Prior Informed Consent (PIC), Mutually Agreed Terms (MATs) and benefit-sharing, without further specification of the instruments and the processes. The Bonn Guidelines provide for very useful guidance, but it is the sovereign States which are called to flesh out the access and benefit-sharing provision of the CBD and adjust them to the local particularities.

The ultimate goal of access and benefit-sharing rules governing bioprospecting activities should actually be financing biodiversity conservation activities and promoting development of developing countries with abundant biogenetic resources. Indeed, developing nations face a variety of problems, such as a population explosion and the need to pay foreign debt and attain economic self-sufficiency. To cope with these pressures their people try to ameliorate agricultural output by seeking more arable land, leading to unchecked deforestation, which is the largest threat to biodiversity and the extinction of species. Thus, with economic pressure being the main cause of biodiversity loss, economic incentives and market instruments are some of the best answers to counterbalance the economic forces leading to the destruction of natural habitats and spur biodiversity

[15] FAO, Commission on Genetic Resources for Food and Agriculture, Animal Genetic Resources at: http://www.fao.org/ag/cgrfa/AnGR.htm (09/09/2008).

conservation.[16] Bioprospecting has been touted by prominent scientists, scholars and international organizations[17] as a mechanism both for discovering new medicines and inducing biodiversity conservation by making it commercially attractive.[18]

Mindful of the fact that reasonable compensation to the providing countries for the production of health-saving medicines creates incentives for the sustainable management and conservation of biodiversity and enhances global welfare, there is a great challenge in this book to explore the implementation options for access and benefit-sharing and to look more comprehensively at the interactions of the main players, in order to evaluate which strategy best serves the CBD's objectives.

## II. METHODOLOGICAL APPROACH

Building on the importance of the bioprospecting agreements as incentivizing tools for the biodiversity protection, it is here supported that to reach this goal it is not merely sufficient that the country that provides for the genetic resources benefits in general; there is also a need to warrant through domestic and international rules and contractual clauses that benefits deriving from the commercialization of genetic resources are channelled to the individuals/communities who actually bear the responsibility of the conservation of these resources. Also, it should be clear from the start that the scope of the current study is limited to plant genetic resources.

[16] See also C Zerner (ed) *People, Plants, and Justice: The Politics of Nature Conservation* (New York, Columbia University Press, 2000) with case studies that offer useful insights for the integration of justice, power, markets and the politics of resource and territorial control in conservation projects.

[17] A number of OECD publications support the argument: OECD *Handbook of Market Creation for Biodiversity—Issues in Implementation* (OECD, Paris, 2004); OECD, *Harnessing Markets for Biodiversity—Towards Conservation and Sustainable Use* (OECD, Paris, 2003); OECD, *Handbook of Biodiversity Valuation—A Guide for Policy Makers* (OECD, Paris, 2002); OECD, *Valuation of Biodiversity Benefits—Selected Studies* (OECD, Paris, 2001); OECD, *Handbook of Incentive Measures for Biodiversity—Design and Implementation* (OECD, Paris, 1999). See also Reid et al (n 1) 17; EO Wilson, *The Diversity of Life* (Allen Lane, London, 1993) 215; and the UNCTAD Biotrade Initiative at http://www.biotrade.org/ (09/09/08).

[18] This idea to finance conservation activities by biodiversity prospecting originates in the pharmaceutical industry: on 22 April 1992, Bill Clinton, while a candidate for American presidency, supported in his speech for the Earth Day a fund of the industrial nations, together with the pharmaceutical industry, to gain access to the rain forest for pharmaceutical research, and make a contribution to the conservation of nature. The proposal was to deposit a part of the profit made by the pharmaceutical industry in the 'Biotic Exploration Fund' to support conservation activities, but such a fund was never established. See U Baruffol, 'Contractual Regulation of Access to Information on Biodiversity for Scientific and Commercial Use—The Novartis-UZACHI Biolead Project' (Diploma Thesis, Forstwissenschaftliche Beiträge 30, Professur Forstpolitik und Forstökonomie, Zürich 2003) 2 (n 2), published on the Internet at: http://www.biodiv.org/doc/case-studies/abs/cs-abs-mx-uzachi-en.pdf (09/09/08).

In this first chapter, a special section is dedicated to the definition and valuation of biodiversity, which justifies incentives for conservation. It is intended to show that because of the special character of biodiversity, biodiversity valuation is a very difficult task, and this renders access and benefit-sharing regulatory design complicated. In addition, a brief overview of the two major industries for the development of medicines and of the process of drug development is provided, as the necessary context in the study of implementation options for access and benefit sharing in this sector from a policy and legal perspective.

Chapter Two commences with the legal treatment of genetic resources before the CBD. Genetic resources were first handled as commons, and the benefits from the intellectual property tools of their derivatives never reached the stakeholders in the source country. Now, the CBD pronounces genetic resources as common concern of mankind, and opts clearly for the sovereignty of the States over the genetic resources within their territory. Then, whether genetic resources belong to the commons or whether they are defined as self-contained objects of legal protection rests on the discretionary power of the States. Because the CBD is quite vague in assigning obligations to the States, the chapter continues with the analysis of the three CBD objectives, to better grasp their meaning and detect implementation guidelines. It is observed in the end that the three objectives of the CBD are not separate, but closely related, and more specific tasks are intended to be drawn from this analysis at the domestic and international levels.

Chapter Three explores the access and benefit-sharing provisions within the CBD, and presents the main features necessary in any access and benefit-sharing legislation. The activities of the CBD Conferences of the Parties are then described, focusing on the Bonn Guidelines, the most elaborate international non-binding instrument today on access and benefit-sharing. Finally, the implementation options for access and benefit-sharing are displayed, namely international instruments, domestic legislation and private practice; the implementation options are examined in a more detailed manner in the following chapters.

In conjunction with Chapter Three, Chapter Four illustrates the activities in place in two other relevant organizations, the WTO and the WIPO. The FAO's multilateral system set in place for the plant genetic resources for food and agriculture is also explored, in order to draw parallels between this system and the bilateral one established by the CBD. All these activities in international fora aim to reconcile the diverse interests among the actors in the international arena and question the possibility that a one-size-fits-all international treaty on access and benefit-sharing will be soon realized.

The primary responsibility for the materialization of the CBD provisions falls upon the domestic legislator. Chapter Five is devoted completely to the domestic implementation option, specifically describing and scrutinizing the Philippines and the Andean Pact regime, because they have the older and most elaborate national and regional, respectively, access and benefit-sharing regulation.

Chapter Six studies the private access and benefit-sharing policies on access and benefit-sharing. At present, many private corporations set policies in their activities with regards to human rights, labor and the environment. With regard to bioprospecting, corporate policies of the drug industry, institutional policies and professional societies' codes of conduct and guidelines complement each other and contribute to the shaping of the whole access and benefit-sharing framework.

Finally, Chapter Seven makes a final evaluation of the present state of play. Following the normative analysis in the second chapter and the observations made from the consecutive study of the international, domestic and private implementation, it is concluded that the most effective and least costly way to achieve the objectives of 'conservation', 'sustainable use' and 'fair and equitable benefit-sharing' is to harness the creative power of entrepreneurs, respecting their indisputable freedom to interchange and direct it to those responsible for quality of biodiversity. For this reason, the focus in this chapter is placed on property and indigenous rights; access and benefit-sharing policies must be accompanied by clear property frameworks and respect for indigenous rights. Given the obstacles for an international access and benefit-sharing treaty, and the difficulties of the countries with users to regulate the international behaviour of these multinational actors according to international law, the burden for the implementation of the access and benefit-sharing call of the CBD falls upon the provider country. But, here, the role of the State should be restricted in generating the institutions to monitor market procedures and the stakeholders' rights, thus directing private expenditure towards biodiversity support, while the advantages of the regional approach are highlighted.

## III. TERMS AND DEFINITIONS[19]

### A. *What is Biodiversity?*

Biodiversity is a fuzzy concept; both its definition and its valuation depend on the people using it. The term 'biodiversity' was first coined in 1986 by two American scientists, Walter Rosen and Edward Wilson, in the preparation of the later proceedings of the first 'National Forum on BioDiversity'.[20]

[19] Previous versions of these sections were published in A Smagadi, 'The Impact of the Convention on Biological Diversity on the Utilization of Plant Genetic Resources and Benefit Sharing (2006) 6 Yearbook of European Environmental Law 120.

[20] S Vermeulen, *Biodiversity Planning: Why and How Should Local Opinions Matter?*

The speed of biodiversity loss is increasing, and this implies disastrous consequences for the future of the biological chain.[21] The crisis of species depletion encompasses the quick extinction of plant and animal varieties and is accompanied by genetic erosion, as their genes disappear forever. This results in the genetic uniformity of species, and constitutes one of the major global problems of our times.

According to the CBD Article 2, '*biological diversity* means the variability among living organisms from all sources including, *inter alia*, terrestrial, marine and other aquatic ecosystems and the ecological complexes of which they are part; this includes diversity within species, between species and of ecosystems'. Very briefly the term describes the diversity of life-forms on this planet.

What really matters in the term 'biodiversity' is not merely the 'ecosystems' and the 'organisms', but the 'diversity'.[22] 'Diversity' stands for variability and variety, that is the living world's capacity to change, and the range of biological forms and processes that derive from this. The interactions of the different components of ecosystems supply essential ecosystem services, and also offer social and recreational opportunities and are a source of inspiration and cultural identity. Biodiversity is not always 'good'; the genetic distinctiveness of the human immunodeficiency virus (HIV) or the boundless proliferation of fungal crop diseases, are undesirable biodiversity expansions.

Thus, biodiversity has a number of distinguishing features that differentiate it from other conventional management resources and are responsible for its complexity. It is a renewable natural resource, if it is used in a sustainable way, but it is definitely not renewable if it is not used in a sustainable fashion. Once a species has completely disappeared, its extinction is ultimate, and belated sustainable use methods do not renew the resource.

Finally, 'biodiversity' does not merely depict the variety of animals, plants and ecosystems. Besides this material aspect of the concept, 'biodiversity' further stands for the variability of life, the diversity of all the living organisms and their interdependence, and the evolutionary capacity of the system through space and time

(Gatekeeper Series, No 115, International Institute for Environment and Development (IIED), Natural Resources Group and Sustainable Agriculture and Rural Livelihoods Programme, London, 2004, published on the Internet at: http://www.eldis.org/cf/rdr/rdr.cfm?doc=DOC15506 (12/09/08).

21 Wilson (n 17).

22 See also OECD (2002) (n 17) 24–26.

*Table 2: Traditional categorization of the types of goods*

| | | **Exclusion** Yes | No |
|---|---|---|---|
| **Rivalry** | Yes | **Private Good** (eg personal computers, bread) | **'Mixed' Good** (Common Pool Resources – CPR) (public libraries, irrigation systems, goods subject to congestion and depletion, but accessible to all) |
| | No | **Toll Good** (eg leisure clubs in science/national parks/television transmission) | **Pure Public Good** (global commons) |

### *B. Environmental Economics and Biodiversity Conservation*

The economics of natural resources and the environment together with property rights theory offer useful insights on why and how the market can be employed for conservation purposes.[23]

A resource such as biodiversity generally makes a multiplicity of services and goods available. The Theory of Goods distinguishes between several types of goods, which are determined by two basic characteristics: the possibility of exclusion of other consumers and the rivalry between the consumers. The above table illustrates the different types of goods according to the two criteria of rivalry and exclusion.[24]

The character of a good can change over time or by setting up or changing property rights. The contributions of Coase (1960) and Demsetz (1967) suggest that property rights will come to be established either de facto in the form of contracts between parties or de jure when the benefits of defining property rights exceed the costs of their enforcement.[25]

In a market economy, where multiple decisions on the patterns of production and consumption relating to goods and resources are made, these decisions usually, but not necessarily, cause costs or benefits to stake-

[23] The economic literature on the subject is vast; see for instance E Kula, *Economics of Natural Resources and The Environment* (Chapman & Hall, London, 1992) and DW Broomley, *Environment and Economy: Property Rights & Public Policy* (Basil Blackwell, Inc, Cambridge, Mass, 1991).

[24] See also Baruffol (n 18) 8 drawing the table from I Kissling and T Volken, *Ressourcenökonomie III. Institutionelle Aspekte der Ressourcennutzung, Unterlagen zum Fachgebiet Forstliche Ressourcenökonomie*, Professur für Forstpolitik und Forstökonomie, Departement Forstwissenschaften ETH, Zürich, 2000, 18.

[25] T Merrill, 'Introduction: The Demsetz Thesis and the Evolution of Property Rights' (2002) 31 Journal of Legal Studies S331; H Demsetz, 'Toward a Theory of Property Rights II: The Competition between Private and Collective Ownership Rights' (2002) 31 Journal of Legal Studies S653.

holders other than the persons making them. In other words they cause externalities.[26] This means that the decision-maker does not bear all of the costs (negative externalities) or reap all of the gains (positive externalities) from his or her actions.

The failure of the market to internalize these externalities is commonly known as 'market failure'.[27] Many environmental problems are translated into economic terms as externalities brought about by market activities. By way of illustration, the production/consumption of a good/resource can produce pollution, a negative externality in environmental economics terminology, which harms not only those in the region of the production/consumption process, but also all humanity by causing global warming effects and climate change.

To mitigate social costs and scarcity of resources, the government plays an important role in correcting market failures with a series of measures that aim at the internalization of externalities and range from property rights to more interventionist public policies.[28] The government's role in tackling market failures is most limited in the so-called '*Coase Theorem*'[29] case. According to Ronald Coase, individuals/groups could organize negotiations/bargains, eliminate externalities, and produce efficient outcomes, even without governmental intervention as long as all three following conditions are met: property rights are well defined, people act rationally, and transaction costs are minimal. The government is the facilitator of negotiating/bargaining process among the affected individuals/ groups, by delimiting property rights and enforcing resulting contracts.[30] However, if a democratic society has other traditional forms of property institutions and contract enforcement evolved over the years to deal with such externalities, the government role might be even cancelled.

Biogenetic resources have long been non-rival and non-exclusive goods, because they were considered to belong to the global commons. Until the CBD expressly enunciated State sovereignty over genetic resources in 1992, the leading opinion in international scholarship considered genetic resources part of the common heritage of mankind as a customary law

[26] See OECD (2004) 35-36; OECD (2003) (n 17); EU Petersmann, *International and European Trade and Environmental Law after the Uruguay Round* (Kluwer Law International, London, The Hague, Boston, 1995) 11–17.

[27] Market failure is the private sector analogy of the less studied government failure: a government failure arises, when the government fails to allocate efficiently goods/resources to government consumers, eg subsidies encouraging farmers to use fields which need intense fertilizers use and irrigation damage the environment (negative externality).

[28] Petersmann 13 Table 4 on the 'Protection of the Environment as a Problem of Private and Public, National and International Law' Petersmann (n 25).

[29] RH Coase, 'The Problem of Social Cost' (1960) 3 Journal of Law and Economics 1.

[30] ibid.

development, but not everybody embraced this view.[31] The adoption of the sovereignty principle in the CBD clarified this controversy in the international legal treatment of genetic resources inclining to a Coasean/ Demsetzian solution to the problem of internalization of the created externalities from the use of biodiversity.[32]

More specifically, global commons are free to be used by anybody and the free use of biogenetic resources in the production of medicines causes a market failure ('biopiracy') because the bioprospecting market fails adequately to involve and compensate all stakeholders, directing them to alternative and more destructive uses of biodiversity; or because the sourcing of the biological resources to be used in the bioprospecting undertaking is unsustainable and threatens species with extinction.[33] Given that delimitation of property rights throughout the world cannot be regulated by an international treaty, the least that the CBD could have done in this case was to assign sovereignty rights to the States. States bear the international obligation to ensure the conservation and sustainable use of the resources, while at the same time the right to profit from their use. The determination of legal and institutional treatment of genetic resources within the state boundaries will have important ramifications in the efficiency of bioprospecting undertakings, implying that effective implementation of the CBD objectives depends heavily on the States' biodiversity policy and more particularly on their access and benefit-sharing policy.

### *C. Biodiversity Valuation*

Biodiversity valuation is necessary so as to find out whether or not biodiversity preservation is important. The importance of biodiversity for human beings, and hence the justification of its conservation, rests exactly on its variety and variability, on the multitude of choices it offers for present and prospective uses and benefits.

Biodiversity has many benefits, but there is no consensus among either

[31] More analytically on the legal treatment of genetic resources in international law, see Chapter 2 Section II.

[32] AC Kiss, 'La notion de patrimoine commun de l'humanité' (1983) 175 *Recueil des Cours—Collected Courses of the Hague Academy of International Law, Académie de droit international* 103; DR Simpson, RA Sedjo, and JW Reid, 'Valuing Biodiversity for Use in Pharmaceutical Research' (1996) 104 The Journal of Political Economy 166.

[33] See J Goodman and V Walsh, *The Story of Taxol: Nature and Politics in the Pursuit of an Anti-Cancer Drug* (Cambridge University Press, 2001); K Ten Kate and A Wells, 'The Access and Benefit-Sharing Policies of the United States National Cancer Institute—A Comparative Account of the Discovery and Development of the Drugs Calanolide and Topotecan' Submission to the Executive Secretary of the Convention on Biological Diversity by the Royal Botanic Gardens, Kew, 1998.

*Table 3: Biodiversity values*

| | |
|---|---|
| **Direct use value** | Biodiversity which can be directly consumed, traded or used for commercial and recreation activities, eg plants or animals collected or hunted for nutrition, clothing or housing, and also sight-seeing and eco-tourism, etc. |
| **Indirect use value** | Functions of the biodiversity which contributes to the well-being of humans, eg natural flood control, prevention of soil erosion or purification of water supplies, habitat, storage and recycling of human produced wastes, etc. |
| **Option value** | The option for direct or indirect use of biodiversity in the future, according to existing knowledge on the use, eg natural gas. |
| **Quasi option value** | The option for direct or indirect use of biodiversity in the future, according to information still unknown on the use, eg stratospheric ozone layer had no known use value before the discovery of its capability to absorb UV-radiation. |
| **Existence value** | The value of biodiversity for its pure existence. |
| **Bequest value** | The value placed on biodiversity maintenance for future generations. |

biologists or economists as to its accurate measurement.[34] The OECD distinguished the following values of biodiversity,[35] exemplified in the table above.

To elaborate,[36] direct use values involve elements of biodiversity which can be directly consumed, traded or used as input for commercial activities, such as plants or animals collected or hunted for nutrition, clothing or housing, including amenity values, such as sight-seeing and eco-tourism. Use values can be realized by individuals and/or companies and are privately appropriable, while market prices are indicative for the direct use values' assessment. Clear property and use rights are indispensable for the realization of use values; ill-defined property and use rights lead to careless short-term exploitation or sub-optimal uses, such as unlicensed timber-harvesting,

[34] DR Simpson, *Definitions of Biodiversity and Measures of its Value, Resources for the Future* (Resources for the Future Discussion Paper 02-62, Washington, DC, 2002) published on the Internet at: http://www.rff.org/rff/Documents/RFF-DP-02-62.pdf (12/09/08).

[35] OECD (2004) 63–103; OECD (2002); OECD (2001); OECD (1999) 27–31. For All OECD (n 17)

[36] This section draws on OECD (1999) 27–31.

and ultimately to rapid disintegration of the intrinsic value of biodiversity. Hence, well-defined property rights in connection with the provision of information about long-run impacts constitute important incentive measures for the realization of use values.

Indirect use values or ecosystems services are all functions of the environment which afford a healthy environment and provide direct value to the well-being of humans on a local, regional or global level; this is the functioning property of the environment and includes natural flood control, prevention of soil erosion or purification of water supplies. The value of ecosystem services can be assessed by calculating the amount which is necessary to invest in installations or technologies to replace ecosystem services. For example, the value of a healthy ecosystem assuring high water quality can be depicted in the price of a water purification plant that replaces it. Unlike the case of direct use values, the benefits arising from the indirect biodiversity values are not appropriable; they have a public/social dimension, and incentives are given through institutional mechanisms (for the assessment of the services provided) and fiscal measures (emission taxes).

Option and quasi-option values describe the values contained in the ability to make choices in an uncertain future. Option values concern choices that people would like to be able to make if their preferences change, whereas quasi-option values concern maintaining the ability to react to future information independently of one's own current preferences and knowledge. For instance, people did not attach use value to natural gas, because they had focused on other energy resources, but it already contained an option value, which is now attached, because other sources of energy are diminishing and the preferences have changed (option value); or the stratospheric ozone layer had no known use value before the discovery of its UV-radiation absorbing capacity, but there was a possibility that it might have some yet unknown use value (quasi-option value).

In the biodiversity context, the option value represents the possibilities of the people to change their preferences in relation only to the biogenetic resources existing in the future, whereas the quasi-option value encompasses not only the biological resources, but also the associated information, which is not available at the moment. Access and benefit-sharing agreements can serve as good instruments to capture the option and quasi-option values of biodiversity.

Some resources provide satisfaction by their mere existence, even if we are not aware of their use, and existence value means that humans value ecosystems and biological diversity for their pure existence. Existence values are very close to cultural and religious values and belong to peoples' general conceptions of nature; the way people perceive existence values depends on the social standards, while existence values increase with income levels. Groups experiencing existence values are subsidizing groups which have not realized them yet through funds, grants, non-governmental organizations, etc. Finally, biodiversity bequest value expresses the value it has for the possibility to maintain it for future generations.

A different categorization includes the commercial, ecological, cultural, esthetical, ethical and intrinsic values,[37] reflecting the same categorization demonstrated above, though with different wording. Agricultural and pharmaceutical products (food and medicine), energy, and raw material, represent the commercial value of biodiversity (direct use), while biodiversity has an essential ecological value by providing substantial benefits for our living, such as environmental services, carbon sinks, and watershed protection (indirect use). Biodiversity has a cultural value in terms of its role in our shared sense of history and place in the world, whereas the aesthetical as well as ethical values should not be underestimated (non-use values). Finally, the intrinsic value expresses the right of non-human species to existence.

Moreover, moral philosophy offers useful insights for biodiversity preservation based on consequentialist, deontological and 'virtue ethical' thinking.[38]

Consequentialist thinking, for example utilitarian thought in public policy inquiries, relates to the effects of given states of affairs. A consequentialist/utilitarian is prepared to weigh the competing interests of different policies and as a result he is ready to accept trade-offs or to engage in cost–benefit analysis. Economics can be grounded on a consequential/utilitarian approach. The utilitarian looks for the choice that produces the most happiness in the society and weighs the costs and benefits to reach an effective collective choice.

Deontological, or non-consequentialist, thinking is the approach that considers certain values as foundational and detached. In contrast to consequential/utilitarian/economic analysis, a deontologist will not accept to weigh different values with intrinsic force, and more importantly, he does not evaluate actions in relation to the results they produce; actually such an approach is unacceptable to him. For example, to an ecologist or an NGO, usually representing the deontological approach, it is morally or ethically wrong to destroy a rare plant species, regardless of the consequences.

Finally, 'virtue ethics' is a revival of the Aristotelian approach to ethics, looking at properties of character, such as courage, temperance, prudence and justice; a virtuous person would try to ask whether biodiversity conservation furthers a virtue of justice or prudence.

37 Vermeulen (n 20) 4.

38 J Linarelli, 'Treaty Governance, Intellectual Property and Biodiversity' (2004) 6 Environmental Law Review 21–38.

The CBD itself does not decide on one of the above rationales. Its objectives, namely use maximization, and equity and fairness in benefit-sharing, create a good amount of uncertainty. In the CBD text, we find strands of all these approaches: in the preamble the intrinsic value of biodiversity is acknowledged (deontological approach) together with the 'ecological, genetic, social, economic, scientific, educational, cultural, recreational and aesthetic values' (consequentialist thinking). Further, exploring concepts of justice in determining sharing allocation suggests virtue ethics reasoning.

To sum up, biodiversity presents a variety of values, which speak for its conservation and sustainable use. Actually, the question of biodiversity valuation is of political, legal and economic nature. In any case, exclusive focus on economic terms can insufficiently represent the complex and multidimensional interrelationship of species in an ecosystem, because economic valuation criteria tend to be fluid, shifting and opportunistic in their practical application; there is a risk that economic analysis alone is not able to embrace the value systems needed to conserve biodiversity over the course of decades and centuries and may lead to wrong results.[39]

## IV. BRIEF OVERVIEW OF THE TWO INDUSTRIES

The distinction between pharmaceutical drugs and botanical medicines is central, not only because the R&D processes used in the two industries are distinct, but also because genetic resources and traditional knowledge are employed in a different manner. Understanding the structure and market of the pharmaceutical and botanical drug industries, helps in realizing the characteristics and needs of these main actors in the bioprospecting undertakings and yields useful insights on the efforts to design access and benefit-sharing regulation. Decisions relating to R&D are both scientific and economic; science delineates the opportunities and the obstacles, but the final decisions in the private sector depend on economic assessments of the financial gains to be harvested from the investments. The basic research is often initiated by governments and conducted by universities and government agencies; however, the great bulk of drug R&D is principally financed by the private sector, which plays an important role in the drug R&D decision-making.

### A. *Pharmaceutical Industry*

The pharmaceutical industry is the most research- and risk-intensive industry in the world. The investments made for the pharmaceutical R&D are

39 Baruffol (n 18) 9 citing Ehrenfeld.

very high and the time needed usually very long, taking on average 15 years, due to increased research and regulatory complexity. Most of this time is spent at the clinical evaluation level of the drug development, which must meet increasingly rigorous requirements. The risk is large because the probability of success is low and the investment returns are equally low, as only approximately 30 per cent of the new chemical entities (in other words of the chemical molecules developed by a company in the very early stages of drug development process)[40] make profit. Pharmaceutical companies invest a higher proportion of sales in R&D than most other industries. The Pharmaceutical Research and Manufacturers of America (PhRMA) reports that in 2003, PhRMA member companies invested an estimated $33.2 billion on research to develop new treatments for diseases—an estimated 17.7 percent of domestic sales on R&D—a higher R&D to sales ratio than any other US industry.[41] At the same time, the prices for the products that reach the market are also high, and the industry is one of the most profitable and one with the most political influence. After all, it was the lobbying from the pharmaceutical industry, in tandem with that of the music industry, that drove to the conclusion of the TRIPS Agreement in 1994.

It was estimated that more than half of all prescriptions filled in the USA in 1993 contained at least one major active compound derived from or modeled on compounds sourced from biological resources.[42]

The biggest market for pharmaceuticals is the US, followed by Japan and then Germany and the UK. The US Pharmaceutical Industry investment has increased from US$ 1.3 billion in 1977 to an estimated US$ 32 billion in 2002.[43] Although the pharmaceutical industry remains one of the best performing high-technology sectors in Europe, Europe is losing competitiveness compared to the US,[44] because most European governments employ a variety of mechanisms to artificially control the prices of prescription drugs; similar governmental price control is observed in Japan. For this reason the concentration of pharmaceutical R&D in North America is rather high.[45]

In developing countries multinational companies hold the majority of the local market, unless the national policy is local manufacture instead as a

[40] See US Federal Food, Drug and Cosmetic Act Section 505(b).

[41] The Pharmaceutical Research and Manufacturers of America, *Pharmaceutical Industry Profile 2003* (PhRMA, Washington DC, 2003) 10 f. PhRMA represents the US main research-based pharmaceutical and biotechnology companies, see www.phrma.org.

[42] F Grifo et al, 'The Origins of Prescription Drugs' in F Grifo and J Rosenthal (eds) *Biodiversity and Human Health* (Island Press, Washington DC and Covelo California, 1997) 131 f.

[43] Thirty-six per cent of pharmaceutical R&D worldwide is performed in the USA, 84 per cent of which is oriented towards new chemical entities; see PhRMA (n 41) 10 f.

[44] European Federation of Pharmaceutical Industries and Associations, *We research for people's health-Annual Report 2001–2002* (EFPIA, Brussels, 2002) 3 and 15.

[45] PhRMA (n 41) 10 f; EFPIA (n 44) 15 f.

form of import-substitution. In some, such as China, Egypt, India and Brazil, the national policy financially supports the local manufacture of generic drugs. Among them, India is considered the most vibrant, generic and low-cost pharmaceutical industry in the world and it indigenously meets around 70 per cent of the country's demand for pharmaceutical intermediate bulk drugs, drugs intermediate and pharmaceutical formulations.[46]

The major players in the pharmaceutical drug industry are private companies, large and small, public sector research product control and marketing agencies, regulatory authorities, and university and academic institutions.

The bulk of pharmaceutical development is conducted by the private sector. There are few companies that produce pharmaceuticals alone, known as 'pharmaceutical houses'. Usually they manufacture and sell a combination of nutritional, medical, agricultural, or chemical products, cosmetics, and vitamins.

A distinctive feature of the pharmaceutical industry is the existence of 'core' companies holding almost one third of the world sales. Over the past decade the industry was reshaped through a series of mergers and acquisitions. Parallel to this consolidation, there is also a trend for re-organization of the pharmaceutical drug industry by knowledge-/science-intensive small and medium-sized enterprises (SMEs) or dedicated biotechnology firms.[47]

Small companies usually specialize in one stage of drug development. They are innovative, more flexible and easier to operate, but a drawback is their difficulty in entering the market, because of the dominating position of the pharmaceutical giants.

Public sector research produces results that become the basis for pharmaceutical development in the private sector. In the context of basic health policy, governmental agencies and non-profit research institutions conduct research, or provide funding for product research or further development, and promulgate and administer regulations for the approval and circulation in the market of a pharmaceutical product. The National Cancer Institute (NCI), one of the seventeen US National Institutes of Health (NIH) in the US, has played a momentous role in the preclinical anti-tumour and anti-HIV substance screening over the past few years. Once public sector agencies come up with a bioactive compound, they are not allowed to market the product, but a private sector institution is found to continue the development and to penetrate the market.

[46] K Ten Kate and S Laird, *The Commercial Use of Biodiversity—Access to Genetic Resources and Benefit-Sharing* (Earthscan Publications Ltd, London, 1999) 35.

[47] P Gehl Sampath, *Regulating Bioprospecting: Institutions for Drug Research, Access and Benefit-Sharing* (UNU Press, Tokyo/New York, 2005) 17 f; ETC, Oligopoly, Inc. *Concentration in Corporate Power: 2003*, ETC *Communiqué*, Issue 82, November/December 2003.

Public research agencies and private companies often form partnerships with academic institutions. The projects carried out by the International Cooperative Biodiversity Groups always involve university or other research institutions, both in the user and the source country, and are always coordinated by US academic institutions. The Costa Rica National Institute of Biodiversity (INBio) that in collaboration with local universities, museums, and government agencies has developed a role as collector, manager, and supplier of genetic resources, their derivatives and information associated with these resources is indicative of this constellation.

To a large extent, combinatorial chemistry is the basis for drug R&D, but even if the artificial synthesis of natural compounds is possible, natural products still keep their place in R&D. Not only the use of natural resources may be proven a cheaper alternative than developing infrastructure to produce the genetic material through other means, but ultimately it does not make sense from an economic point of view for companies to invest in developing alternative sources of supply, until it is evident that a commercially viable drug can be developed from a specific compound; and the degree of certainty is rather low.

### *B. Botanical Industry*

The botanical industry is experiencing rapid growth worldwide, ranging from 10–20 per cent in most countries, while the major markets are Europe (Germany leading), USA, Japan, and China.[48]

Botanical (or herbal) medicines or phytomedicines are part of larger markets referred to as the 'dietary supplements' market in the US for instance (including vitamins, minerals, herbs/botanicals, and other natural medicines), often joined by nutraceuticals (foods with health-enhancing properties). All these different botanical market products are sold at different levels of processing and in different forms, such as dried and fresh herbs, tinctures, extracts and herbal teas, capsules and tablets. When these products are standardized active compounds in an herbal base, they are known as phytomedicines.

The increasing interest in botanical medicines is due to their 'proven' efficacy and safety and their easy and fast manufacture. The trend among the Western consumers is to show a preference for botanical to pharmaceutical

[48] Centre for the Promotion of Imports from Developing Countries, *EU Market Survey 2005: Natural Ingredients for Pharmaceuticals* (CBI, The Netherlands, 2003); Association Européenne des Spécialités Pharmaceutiques Grand Public, *Herbal Medicinal Products in the European Union—Study carried out on behalf of the European Commission* (AESGP, Brussels, 1998);Ten Kate and Laird (n 46) 78 f; CBD Submission by the United Nations Environment Programme *Benefit-Sharing Case Studies: Aristocladus korupensis and Prunus africana*, Doc. UNEP/CBD/COP/4/Inf.25, 20 April 1998, 31.

drugs. Traditional knowledge is an important complement to scientific testing and is also very broadly used for marketing purposes. In the extensive study carried by ten Kate and Laird,[49] all the companies interviewed made use of documented traditional knowledge associated with the species.

Thanks to scientific advances, the species of St John's Wort (the number one anti-depressant medication in Germany since 1995, when it outperformed the pharmaceutical Prozac), ginseng, gingko, ginger, kava, green tea, and garlic feature as major products in both US and European markets.[50]

The global botanical medicines market is very diverse, reflecting the scientific, economic and cultural roots of the national industries. In the US, the growing market in the 1980s was still limited to people selling herbs to each other and was based on the knowledge and practices of the US's indigenous peoples. In contrast, in Europe and especially in Asia, there is a longer tradition. In particular in China and India, botanical medicine markets are current expressions of formal medical systems developed over thousands of years. Large pharmaceutical companies have lately entered this business and spurred the expansion of the industry into a mass market.

The main characteristics of the rapidly developing botanical industry are the consolidation of distribution channels and vertical integration, and the sale of botanical medicines in a wide range of outlets, especially mass market outlets like supermarkets and drugstores. The structure and nature of participation in the market companies varies. Company size, function (in cultivating, supplying, manufacturing, and distributing), philosophy and strategy vary widely.

Expanded research has improved the legitimacy of botanical medicines. The regulatory environment is not friendly everywhere, but this is changing. In the US, the development of botanical drugs has for many years demanded the combination of the merits of advanced Western technology and the empirical-based research and the concept of botanical drugs has placed herbal medicine into the demanding FDA approval process. However, while natural substances cannot be patented, companies can claim exclusive rights for processes and this can be an incentive to invest on that side of product development. Recently, this regulatory flux in the US has started to change; the manufacture and marketing of botanical medicines are becoming more attractive, while national and commercial insurance companies have started to accept them.[51]

[49] Ten Kate and Laird (n 46).

[50] Among the top retail botanical drugs in Europe are Tebonin (Gingko biloba) by Schwabe (Germany), Ginsana (Ginseng) by Pharmaton (Germany), Kwai (Garlic) by Lichtwer Pharma (Germany), Efamol (Evening primrose) by Scotia (UK), Echinacin (Echinacea) by Madaus (Germany).

[51] W Li, 'Botanical Drugs: A Future for Herbal Medicines' (2002) 19 The Journal of

The situation differs in Europe[52] and especially in Asia, where for many countries botanical medicine has been deeply rooted in the culture and research in the sector is gaining support. In Europe, botanical medicines are comprehensively regulated in most countries, with standards set in terms of safety, efficacy and quality control for manufacturing and the final product, and are integrated into the health care system.

Botanical drugs are steadily gaining consumers, who seek alternative or complementary pharmaceutical drugs and modern healthcare. The general 'green' consumerism in Western societies led to increasing demand for 'natural' medicine, increasing advertising budgets, and media attention.

The sudden sales growth in the botanical industry raises serious benefit-sharing and sustainability concerns. Raw material is exported as a bulk commodity from highly biodiverse countries, access regulations are poor, whereas there are no provisions for sharing of the benefits from the products' commercialization. At the same time, raw material harvesting is not always carried out in a sustainable way, and species have been over-harvested and in danger, because cultivated sources developed for regional or subsistence use cannot satisfy the sudden international demand.

As demand is likely to increase in the coming years, the capacity-building and technology-transfer potential from commercial raw material sourcing

Contemporary Health Law and Policy 117. The 1994 Dietary Supplement Health and Education Act (DSHEA) offered low market cost, but had many drawbacks; the more recent FDA Drug Guidance for Industry on Botanical Drug Products, released in August 2000, signals a step toward more favourable regulatory framework.

[52] See European Parliament, Point 46, Minutes of 16 April 1996, document A4-0104/1996: '[The European Parliament] calls on the Commission to prepare additional proposals on how also to facilitate the European marketing of herbal and homeopathic medicines [...]; requests the Commission to adjust the authorisation procedure to allow such medicines to be generally available throughout the Community [...], calls for the setting up of Traditional Medicines Evaluation Agency, comprising experts in this field, to assess the worth of phytomedicines'; and Commission Directive 1999/83/EC of 8 September 1999 amending the Annex to Council Directive 75/318/EEC on the approximation of the laws of the Member States relating to analytical, pharmacotoxicological and clinical standards and protocols in respect of the testing of medicinal products [2000] OJ L003 P. 0018: ...(6) Whereas it is in particular necessary to clarify that 'bibliographic reference' to other sources of evidence (post-marketing studies, epidemiological studies, studies conducted with similar products, etc.) and not just test and trials may serve as a valid proof of safety and efficacy of a product if an applicant explains and justifies the use of these sources of information satisfactorily...' As regards traditional herbal medicinal products, the European Commission's directive proposal on 'Traditional Herbal Medicinal Products' got integrated in Article 16 of the codified Directive 2001/83/EC of the European Parliament and of the Council of 6 November 2001 on the Community code relating to medicinal products for human use [2001] OJ L 311 P. 0067–0128, and amended with Directive 2004/24/EC of the European Parliament and of the Council of 31 March 2004 amending, as regards traditional herbal medicinal products, Directive 2001/83/EC on the Community code relating to medicinal products for human use [2004] OJ L 136 P. 0085–0090; its provisions are restricted to herbal preparations which are to be included in a specific list to be set up in the near future.

and R&D partnerships is important. Mega-biodiverse countries have a great opportunity to profit from their resources and knowledge. However, the increasing demand of raw material and the simultaneous decreasing wild populations of many species are alarming and restrain the possibilities of the botanical industry. Wild populations of species like Prunus Africana have been harvested in unsustainable ways in order to feed international markets. It is necessary that access and benefit-sharing guidelines are closely tied to sustainable use and other environmental conservation provisions, so as to ensure that close monitoring of the trade in raw materials according to international standards, such as those defined by the Convention on International Trade in Endangered Species (CITES).

## V. DRUG DEVELOPMENT AND BIODIVERSITY

### *A. The Role of Genetic Resources and Traditional Knowledge on Drug Development*

One component of biodiversity is genetic material, that is, 'any material of plant, animal, microbial or other origin containing functional units of heredity'.[53] Genetic resources are the chemical chromosomal information carried in gene alleles of living cells. They furnish the raw materials for plant breeders and biotechnologists; indeed, the diversity of genes and species is of utmost importance to food-related and medical research. Numerous food products, medicines and biotechnological applications have been developed from wild species of plants and animals. If such genes were to disappear, so would the potential for many beneficial techniques, such as options for plants and weeds that might serve to feed the growing world population and medicines for existing and future diseases

As already discussed, the production of medicines is one of the largest markets ever, health concerns have led to research and technology and there continues to be much progress: there are pharmaceutical drugs created in the laboratory and there are drugs derived from natural resources. The latter may be entirely based on natural resources; they may derive from natural resources and be further developed through semi-synthetic modifications; they may be of biological or biopharmaceutical nature, namely proteins and peptides isolated from the natural organic resource's cells, or drugs produced by rDNA techniques; or they may be entirely produced synthetically in the laboratory, but modelled on a natural product.[54]

The production of a medicine may be based on genetic resources from

[53] CBD Article 2.
[54] Ten Kate and Laird (n 46) 34 f and 78 f.

plants,[55] microbial sources,[56] animal genetic resources,[57] marine organisms or insects that lie within the borders of rich in biota developing countries (with the exception of Australia, which is the only mega-biodiverse developing country).[58] Lately, human genetic resources (like human proteins produced by rDNA techniques)[59] have attracted considerable attention as medicinal uses.[60] The Conference of the Parties to the CBD interpreted the access and benefit-sharing provisions of the CBD in such a way as to exclude human genetic resources.[61] However, this aspect of access to genetic resources raises profound ethical questions and remains largely unregulated.

Many companies have natural resource units and conduct research on the medicinal potential of genetic resources. Botanical companies examine and use the naturally occurring compounds in unmodified form, whereas

[55] For the historical significance of plants see MJ Balick and PA Cox, *Plants, People and Culture—The Science of Ethnobotany* (Scientific American Library, New York, 1996). It is estimated that less than 20 per cent of higher plants have been investigated for the presence of bioactive compounds. Also, for the importance of medicinal resources in tropical forests see MJ Balick, E Elizabetsky, and SA Laird (eds) *Medicinal Resources of the Tropical Forest—Biodiversity and its Importance to Human Health* (Columbia University Press, New York, 1996).

[56] E Mathur et al, 'An Overview of Bioprospecting and the Diversa Model' (2004) IP Strategy Today No. 11-2004, 1.; H Doremus, 'Nature, Knowledge and Profit: The Yellowstone Bioprospecting Controversy and the Core Purposes of America's National Parks' (1999) Ecology Law Quarterly 401.

[57] Animals' hormones and metabolites can be the basis for specialized R&D programs, eg American Home Products's Premarin treatment against osteoporosis uses conjugated oestrogens derived from the urine of pregnant horses. Also, indigenous peoples in the neotropics have long used poisonous compounds from venomous animals –snakes, spiders, insects, scorpions, etc.

[58] The International Cooperative Biodiversity Groups (ICBG) award those currently working in several countries in the world to recognize the medicinal potential of the earth's plants, animals and microorganisms with the objective to integrate improvement of human health through drug discovery, incentives for biodiversity conservation and new models of sustainable economic activity, see the ICBG web site on the Internet at: http://www.fic.nih.gov/programs/research_grants/icbg/ (12/09/08).

[59] rDNA stands for recombinant DNA, that is a process of recombining the DNA sequences of two different organisms and producing a different, genetically modified organism.

[60] The Human Genome Project, an international loosely organized and largely government sponsored initiative, is intended to unravel the human genome. Scientists, both from government and private industry, recently completed the landmark mapping of genes in the human body in a development called the Human Genome Project, which opens new doors to further breakthrough research. See A Dunham et al, 'The DNA Sequence and Analysis of Human Chromosome' (2004) 13 Nature 522. Private ventures such as Incyte Pharmaceuticals, Human Genome Sciences and Celera Genomics are running their own private genome projects using a technique known as expressed-sequence tagging of genes. See Oak Ridge National Laboratory, The Human Genome and the Private Sector—A Working Partnership, published on the Internet at: http://www.ornl.gov/sci/techresources/Human_Genome/project/privatesector.shtml (12/09/08); M Eliot, 'Sharing the Glory, Not the Credit' (2001) 291 Science 1189.

[61] Conference of the Parties to the Convention on Biological Diversity, Second Meeting, Decision II/11 on *Access to Genetic Resources*, Jakarta, 6–17 November 1995, 5 October 1996, para. 2.

the pharmaceutical companies are more inclined to developing 'leads'; in other words, finding promising molecules of naturally detected compounds that must be modified to increase efficacy or reduce side effects.

In terms of process, pharmaceutical R&D is lengthy and costly. It involves the acquisition of materials for screening, identification of a disease or of a therapeutic target of interest, methodology development and identification of active agents and chemical structure. It begins with the identification of active lead molecules, where a number of different techniques (phylogenetic, ecological and ethnobotanical surveys) are used for the selection of genetic resources that may be used; followed by detailed bioassays and formulation, and clinical studies which establish safety, efficacy, and the profile of the new drug,[62] which will be approved by the competent national authority and marketed.

Botanical R&D differs from the pharmaceutical one. First of all, the whole botanical industry is based on natural compounds (plant material) and there is no classification between naturally derived and synthesized ones. Secondly, botanical drugs are produced directly from the whole natural plant material, whereas a conventional drug approved by the national drug authority has a single well-characterized active ingredient derived from the natural resource. A botanical drug, by definition, comes in forms of extracts that are composed of multiple natural chemical constituents and active ingredients and its bioactivity is based on their collaborative action rather than on a single, isolated active compound. Thirdly, a botanical industry is principally based on traditional medicinal practices; most botanical medicines have long histories of traditional use, which confirm their safety and efficacy in their approval from the competent drug authorities. [63]

As also discussed in the previous section, botanical R&D is much simpler than pharmaceutical R&D and traditional knowledge input is very significant. The process begins with clinical evaluation of the treatment modalities practiced by traditional healers and/or used by local communities, which is followed by a stage of toxicity studies. Traditional medicinal knowledge provides a strong basis for safety and efficacy. It is actually possible to convert crude plant medicines into standardized phytomedicines in less than two years.

To a certain extent, the benefits modern society gains in terms of health care and longer life expectancy can be ascribed to genetic resources and traditional knowledge, on the one hand, and to innovation in the private sector, on the other. Biodiversity-rich countries, which possess primary

[62] Gehl Sampath (n 47) 20; Ten Kate and Laird (n 46) 49 f.

[63] Gehl Sampath (n 47) 13 and 23; Ten Kate and Laird (n 45) 78; Doc UNEP/CBD/COP/4/Inf.25, 20 April 1998 (n 47) 32.

*Table 4: Plant-based drugs and medicines*[64]

| Drug | Source | Medical Action/Use |
|---|---|---|
| Morphine | *Papaver somniferum* (Opium poppy) | Analgesic |
| Pilocarpine | *Pilocarpus jaborandi* | Parasympathomimetic (used to treat Glaucoma) |
| Quinine | *Cinchona ledgeriana* | Antipyretic, antimalarial |
| Scopolamine | *Datura species* | Sedative |
| Taxol | *Taxus Brevifolia* (Pacific yew) | Antitumour agent (used to treat ovarian cancer) |
| Tubocurarine | *Chondodendron tomentosum* | Muscle relaxant |
| Vinblastine | *Catharanthus roseus* (Rosy periwinkle) | Antitumour agent (used to treat Hodgkins' disease) |
| Vincristine | *Catharanthus roseus* (Rosy periwinkle) | Antitumour agent (used to treat leukemia) |

genetic material, often accompanied by medicinal knowledge of the locals, and industry exploit these resources, and often instructed by traditional uses, invest large sums of money in research and innovation to create new products. As a matter of fact, a large percentage of all drugs and preparations sold in the Western markets derive from extracts of higher plants. The above table is a very short list of some of the drugs whose active ingredients are extracted or derived from plants.

A key enterprise in biological analysis is what scientists call chemical prospecting or biodiversity prospecting or bioprospecting, meaning the search among wild species for new medicines and other useful chemical products.[65] Because bioprospecting depends heavily on classification of species, it is best conducted in tandem with biodiversity research surveys. In order to succeed, investigators must also work in laboratories equipped with advanced facilities, which are usually available only in industrialized countries.

Screening processes vary: with random screening, genetic resources are collected and detected in an indiscriminate manner; with ethnobotanical screening, collection of samples is guided by traditional uses, whereas in the

[64] See the much longer list by L Taylor MD, 'Plant Based Drugs and Medicines', 13 October 2000, published on the Internet at http://www.rain-tree.com/plantdrugs.htm (15/11/08).

[65] WV Reid et al, 'Biodiversity prospecting' in Balick, Elizabetsky and Laird (eds) (n 55) 142.

broad-based taxonomic sampling, research is guided by the existing knowledge and information on the medicinal properties of one species, and other species within the same plant family and genera are sampled.[66] Traditional knowledge plays an important role in the ethnobotanical screening, because it can guide to a high degree the research initiatives.

The following cases illustrate well how benefit-sharing may be used in the production of medicines.[67]

In 1991, the pharmaceutical giant Merck & Co implicitly acknowledged an obligation to compensate for the seed of raw plant germplasm in a much-publicized venture with the government of Costa Rica (through Costa Rica's National Institute of Biodiversity (INBio).[68] Under the agreement, Costa Rica was to provide Merck with plant, animal or microbial samples, and INBio agreed to contribute 10 per cent of the budget and 50 per cent of any royalties to the government's National Park Fund for the conservation of national parks in the country. Merck agreed to provide technical assistance and training to establish drug research capacity in Costa Rica, however, the program's results on any potential drug candidate are unclear.[69]

In 1993, the US National Cancer Institute (NCI)[70] signed an agreement with the Government of Cameroon following the discovery of a potential anti-AIDS plant, *Ancistrocladus korupensis*, found in the Korup forest.[71]

[66] ibid.

[67] See more detailed contract descriptions of the case studies in Chapter 6 Section IV.

[68] The National Biodiversity Institute (INBio—http://www.inbio.ac.cr) was created in 1989 as a non-governmental, non-profit association for private founding members, with the mission to promote a new awareness of the value of biodiversity and thereby achieve its conservation and use to improve life quality. INBio has a formal agreement with the Ministry of Environment & Energy to explore the use of biodiversity and promote the national inventory in government-protected areas. Under this agreement, INBio actively develops biodiversity prospecting in protected wild areas of the country, in participation with national and international academic and private sectors. The agreements specify that 10 per cent of the research budgets and 50 per cent of future royalties will be donated to the Ministry and reinvested in conservation. For a full account of the experience in bioprospecting in Costa Rica see J Cabrera Medaglia, 'Bioprospecting Practices in Practice: A Decade of Experiences at INBio in Costa Rica' (2004) IP Strategy Today No. 11-2004, 27.

[69] G Sampath (n 47) 30; LH Caporale, 'The Merck/INBio Agreement: A Pharmaceutical Company Perspective' in Balick, Elizabetsky, and Laird (eds) (n 55) 137.

[70] The National Cancer Institute (NCI) (www.cancer.gov) is one of the seventeen US National Institutes of Health (NIH) under the auspices of the Department of Health and Human Services of the United States Government. The NCI was established in 1937, with the mission to provide, promote and coordinate cancer related research. On the NCI's policies on access and benefit-sharing, see LA Salicrup, RF Harris, and ML Rohrvaugh, 'Partnerships in Technology Transfer: An Innovative Program to Move Biomedical and Health Technologies from the Laboratory to Worldwide Application' (2005) IP Strategy Today No 12-2005, 1–12, and Ten Kate and Wells (n 32).

[71] Doc UNEP/CBD/COP/4/Inf.25 (n 46); J Mugabe and J Otieno-Odek, 'National Access Regimes: Capacity Building and Policy Reforms' in J Mugabe et al (eds) *Access to Genetic Resources—Strategies for Sharing Benefits* (ACTS Press, Nairobi, 1997) 107.

Since then, the forest has become one of West Africa's largest conservation projects. Under the agreement, the Government of Cameroon will provide the NCI with plant samples without destroying the forest's resources. The NCI's payments will then be used for Korup's community development projects.

On the basis of the medicinal uses of the arogyapaacha plant by the Kani in South India, scientists at the Tropical Botanic Garden and Research Institute in India isolated the plant's active compounds and the anti-stress and anti-fatigue sports drug Jeevani was developed. Patents were filed and the technology was licensed to an Indian pharmaceutical manufacturer, Arya Vaidya Pharmacy, Ltd. To share the benefits with the Kani, a benefit-sharing agreement was established.[72]

Another famous story is the Hoodia case. The San or Bushmen of Southern Africa used hoodia cactus as an appetite-suppressant in the desert. The South African Council for Scientific and Industrial Research (CSIR) received a patent for Hoodia's active ingredient, then licensed to the British Pharmaceutical Pfizer. After a large campaign from the San in South Africa, a benefit-sharing arrangement was concluded between CSIR and the San.[73]

The contractual approach to bioprospecting has an advantage over sole reliance on intellectual property systems, in that costs, benefits and risks can be better distributed between the collecting organization and the country where the biological material was sourced (source country): contractual provisions determine the beneficiaries' portions and allocations, they are flexible at accommodating the parties' will and they can be concluded even in absence of domestic intellectual property laws in the source country. However, the uneven negotiating powers between the developing source country and the multinational pharmaceutical company with strong legal departments, and the incapacity of developing countries to analyze information provided by the industry are some of the approach's shortcomings.

There are several historical reasons why pharmaceutical companies and other research and commercial organizations are increasingly inclined to take part in bioprospecting.[74] In the 1960s and 1970s, pharmaceutical companies phased out the screening of plants on the grounds that the process was too complicated and expensive. Companies turned to new tech-

[72] Ten Kate and SA Laird, 'Bioprospecting Agreements and Benefit-Sharing with Local Communities' in Finger and Schuler (eds) (n 46) 133; RV Anuradha, *Sharing with the Kanis: A case study from Kerala, India*, 1998, case study submitted at the CBD Secretariat, published on the Internet at: http://www.biodiv.org/doc/case-studies/abs/cs-abs-kanis.pdf (12/09/08).

[73] J Wise, 'Hunter-gatherers Win Profit-sharing Deal for Obesity Drug' Bulletin of the World Health Organization 2003, 81 (5) 381. San people or bushmen use a plant extract in drought periods to restrain the feeling of hunger and thirst. See 'Bushmen Victory over Drug Firms' *The Observer* (Sunday 31 March 2002).

[74] See F Grifo and J Rosenthal (eds) *Biodiversity and Human Health* (Island Press, Washington, DC and Covelo, California, 1997).

nologies in microbiology and synthetic chemistry, preferring to rely on human ingenuity because this seemed more 'scientific' than evolved natural chemistry, as well as being less expensive.

In the 1990s the trend changed again: because of advances in technologies, high-volume robot-controlled biological essays allowed large companies to screen up to 50,000 samples a year using only bits of fresh tissue or extract flown to them from any part of the world.[75]

Now, large companies are finding that the middle-term pay-off may not have been as easy as expected and are reducing their drug discovery programmes to go back to genomics and chemical research derivatives.[76] One principal reason is that drug companies worry about whether they will be able to patent the uses of natural products.

Even though drug and food companies are scaling back their research and bioprospecting activities, it is estimated that this is only a temporary trend, because there is such potential in gene diversity and such novelty in nature that is very difficult to replicate in laboratories.[77] The World Health Organization estimates that 80 per cent of people in the developing world rely on traditional medicines based largely on the use of medicinal plants.[78]

The path of developing goods for commercial production from raw genetic resources can sometimes be shortened further by taking clues from the traditional medicinal knowledge and practices of local and indigenous peoples. Genetic resources embody characteristics of the territory in which they are found; they are part and parcel of the indigenous ecosystem and mutually relate, in terms of belonging and use, with the local people who are also part of the ecosystem. This explains why pharmaceutical inventions are increasingly dependent not only on the natural bio-wealth, but also on the related medicinal knowledge of indigenous populations.

[75] Wilson (n 17) 322; J Mugabe et al, 'Managing Access to Genetic Resources' in Mugabe et al (eds) (1997) 5; and WV Reid, 'Technology and Transfer to Genetic Resources' in Mugabe et al (eds) (1997) 56.

[76] WB Chambers, 'Emerging International Rules on the Commercialization of Genetic Resources—The FAO International Plant Genetic Treaty and CBD Bonn Guidelines' (2003) 6 The Journal of World Intellectual Property 314.

[77] In economics, Simpson, Sedjo, and Reid (1996) developed a model to argue that biodiversity prospecting holds out no hope as a meaningful source of finance for biodiversity conservation. See also CB Barett and TJ Lybbert, 'Is Bioprospecting a Viable Strategy for Conserving Tropical Ecosystems?' (2000) 34 Ecological Economics, 293. These conclusions were opposed by G Rausser and A Small, 'Valuing Research Leads: Bioprospecting and the Conservation of Genetic Resources' (2000) 108 The Journal of Political Economy, 173, who suggested that under plausible conditions, the bioprospecting valued of certain genetic resources could be large enough to support market-based conservation of biodiversity; and also K Day-Rubenstein and GB Frisvold, 'Genetic Prospecting and Biodiversity Development Agreements' (2001) 18 Land Use Policy 205, who agree on the limited set of incentives from bioprospecting agreements to reduce uncultivated land conversion and endorse the inappropriateness of these agreements to correct market failures, but they focus on the non-monetary benefits of the bioprospecting contracts as biodiversity conservation strategy.

[78] See WHO Press Release WHO/38, 16 May 2002.

### *B. Value of Undiscovered Drugs and Incentives Problems*

The odds of a single compound becoming a drug after entering the discovery process in the pharmaceutical R&D is estimated at one in 5000–10,000.[79] For example, since 1960, only seven plant-derived anticancer drugs elaborated from the NCI have received FDA approval for commercial production. According to the NCI's Natural Products Branch, each year scientists test 20,000 extracts and 98 per cent of them do not show activity against cancer or AIDS. Since 1986, over 40,000 plant samples have been screened, but thus far only five chemicals showing significant activity against AIDS have been isolated, and they are currently in preclinical development.[80] Before their consideration for clinical trials in humans, these agents must show tolerable levels of toxicity in animal models usually.

At the same time, estimations of the value of undiscovered drugs are generally high. It is striking that of the 119 known pure pharmaceutical compounds used somewhere in the world, 88 were discovered through leads from traditional medicine.[81]

A major problem to define the value of samples for medicinal use is the time lag between the collection and marketing of the product, which is estimated to be 10–15 years. This creates difficulties in the drafting of benefit-sharing agreements between the providers and the users of the genetic resources.

Further, in terms of benefit-sharing arrangements, the assignment of property rights is important for reasons of efficiency and equity; accepting the existence of transaction costs within industry, proper allocation of property rights influences essentially efficient levels of investment at various levels. Hence, poorly constructed property rights systems inhibit not only the valuation of biodiversity, but also the development of bioprospecting activities for further conservation of the biological resources.[82]

[79] Ten Kate and Laird (n 46) 48.

[80] Taxol, vinblastin, vincristine, topotecan (a canptothecin derivative), irinotecan, etoposide, and teniposide-the last two being semisynthetic derivatives or epipodophyllotoxin. See, National Cancer Institute, 'Cancer Facts: Questions and Answers About NCI's Natural Products Branch', National Cancer Institute—National Institutes of Health, Department of Health and Human Services, (fact sheet reviewed on 5/1/00), published on the Internet at: http://cis.nci.nih.gov/fact/7_33.htm (12/03/2004).

[81] It has been estimated that every new drug is worth US$94 million to a private drug company and US$449 to society as a whole. Also that the world's tropical forests contain 375 potential pharmaceuticals of which 48 have been discovered. The calculations show that a complete collection and screening should be worth US$3-4 billion to a private drug company and US$147 billion to the society as a whole see Baruffol (n 18) 10 and Wilson (n 17) 323.

[82] H De Soto, *The Mystery of the Capital: Why Capitalism Triumphs in the West and Fails Everywhere Else* (Bantam Press, New York, 2000); T Swanson and T Göschl, 'Property Rights Issues Involving Plant Genetic Resources: Implications of Ownership for Economic Efficiency' (2000) 32 Ecological Economics 75.

Different economic studies base the valuation and quantification of biodiversity in terms of drug discovery on different assumptions and methodologies. For instance, the one disease/one drug model by Simpson, Sedjo and Reid[83] is based on the finding that the value of marginal species is low and, hence, pharmaceutical prospecting is not a viable method to finance biodiversity conservation. Moreover, Rausser and Small[84] built their one disease/one drug model on the assumption that once a drug is found, no other drug will be needed; they found that genetic resources have a higher marginal value in targeted search structure. Furthermore, the value of biodiversity has also been quantified in terms of conserved land, which is a useful measure considering that land conversion is the main threat to biodiversity conservation.

In economic terms, the marginal value of something represents its price. In the case of bioprospecting, its marginal value lies in the potential contribution of an additional species to the probability that researchers may be able to find what they are looking for. This means that chemical compounds are valuable only if there are few substitutes for them. There are millions of species (animals, plants, micro-organisms), from which useful compounds can be sourced; their marginal value is low either because they are so common as to be redundant or too rare to make a discovery possible.

There are several reasons why genetic resources may be relatively redundant: it may be that the same species may be found over a wide range; that drugs with similar clinical properties can be isolated from different species; or that different therapeutic mechanisms may be effective in treating the same symptoms, because organisms in similar ecological niches often evolve similar chemicals.[85] The point, eventually, is that the larger the set of organisms that can be sampled, the smaller is the marginal value of species. If their value is small, no incentive for conservation exists, and no arguments can be used against the conversion of land to generate other incomes. If their value is high, it might result in unsustainable harvesting and destruction. Either way, biodiversity becomes extinct.[86]

However, as already explained, biodiversity valuation cannot be based only on economic assumptions, because sometimes economics unrealistically simplifies the facts and lacks accuracy in practical terms. Furthermore, recent technological advances are not considered, and a broad range of social benefits deriving from conservation is poorly investigated and remains unknown.

[83] Simpson, Sedjo, and Reid (n 32).

[84] Rausser and Small (n 77).

[85] eg the anticancer drug taxol was primarily produced from the Pacific Yew of western North America, but after unsustainable use and extinction of its primary source, the drug was afterwards produced by other European Yews that demonstrated similar properties; see Goodman and Walsh (2001).

[86] Simpson, Sedjo, and Reid (n 77) 168–169.

Thus, the value of biodiversity continues to be critical for medicinal production. Besides these clear potential benefits, bioprospecting involves great risks. Since most screening facilities are located in the industrialized world, the benefits may never reach the primary source suppliers, or the identification of a valuable plant may create such tremendous demand that expansive harvesting leads to extinction. The challenge is to address these problems and draw the maximum advantages from bioprospecting.

# CHAPTER 2

# *Convention on Biological Diversity—Principles*

## I. INTRODUCTION

In order to evaluate the best way to implement the CBD's requirement for fair and equitable benefit-sharing and pass operational access and benefit-sharing legislation at the international and domestic level, it is necessary first to examine how the CBD objectives of conservation, sustainable use and of fair and equitable benefit-sharing are understood through space and time. This chapter starts with the examination of the status of genetic resources in international law before the CBD and the consequences of this legal environment. It continues with the modest attempt to analyse CBD objectives incorporated in its first article in a multidisciplinary and inter-temporal manner, resulting in three objectives that are not isolated but are to be viewed in relation to each other. Finally, access and benefit-sharing regulation should provide for a fair process and an equitable distribution, whilst policies supporting indigenous peoples' rights and clear property rights promote the CBD objectives.[1]

## II. GENETIC RESOURCES IN INTERNATIONAL LAW: BEFORE THE CBD

The link between genetic resources and intellectual property means that genetic resources may have been altered by human intervention and may have taken on characteristics not found in nature; when those alterations lead to an invention, which is new and applicable on an industrial scale, the invention may qualify for a patent.

Throughout recent history, a variety of conceptual frameworks have been used to regulate the access to and trade of genetic resources. Chronologically there are three main approaches: the common heritage doctrine, international patent law, and finally, the United Nations Convention on Biological Diversity. In this section the two first approaches will be examined.

[1] Prior versions of Sections II and IIIA have been published at A Smagadi, 'The Impact of the Convention on Biological Diversity on the Utilization of Plant Genetic Resources and Benefit-Sharing (2006) 6 Yearbook of European Environmental Law 120.

### A. Initial Assumption: the Common Heritage Principle

Exotic plant species have historically been classified as common heritage of humanity. Due to their classification these resources could be accessed by a large number of peple and used without restriction.[2]

Although naturally occurring plant genetic resources have been treated as subjects of free exchange among nations, industrial countries with intellectual property systems protected plant-derived breeding and bioengineering products developed by their seed and drug companies. Rich in biodiversity, developing countries complained about this exploitation of their genetic resources, arguing that foreign researchers freely prospected in their forests for native landraces and species with medicinal and other functions and then sold expensive, patented products based on the native species back to developing countries (biopiracy).

In 1983, a resolution was signed to highlight the importance of plant genetic resources for food and agriculture and the International Undertaking on Plant Genetic Resources, was adopted by FAO.[3] The resolution aimed at assigning responsibility in all humanity for the conservation of these resources and stated that all plant germplasm was deemed 'common heritage of mankind'. It was the first international agreement to deal with sustainable management of plant genetic resources at the global level.

However, the application of the common heritage doctrine to plant genetic resources as advocated by the International Undertaking proved unsuccessful, because the scope of the traditional definition of the common heritage principle in international law cannot encompass genetic resources that fall within the borders of a specific sovereign nation and are thus treated as national or private property.

The common heritage principle[4] was first enunciated in the Law of the Sea negotiations.[5] Since then it has gained international acceptance in areas such as the deep sea-bed, the lunar surface, and Antarctica, which unlike plant genetic resources fall outside the jurisdiction of any State. This is a critical difference and has decisive consequences in terms of exploitation. Nations can exercise de facto ownership of their natural resources and choose to exploit or preserve their native species as they please.

[2] W Lesser, *Sustainable Use of Genetic Resources under the Convention on Biological Diversity: Exploring Access and Benefit-Sharing* (CABI Publishing, New York, 1997) 13 f.

[3] FAO Conference Resolution 8/83, 22nd Session, International Undertaking on Plant Genetic Resources.

[4] S Bilderbeek (ed) *Biodiversity and International Law—The Effectiveness of International Environmental Law* (IOS Press, Amsterdam/Oxford/Washington, DC/Tokyo, 1992) 86 f.

[5] United Nations Convention on the Law of the Sea (10 Dec 1982) (1982) 21 International Legal Materials 1261, Art 136 containing the first appearance of the common heritage argument in international law.

Another argument speaking against the suitability of the common heritage principle in the case of genetic resources is the so-called 'tragedy of the commons'.[6] The tragedy of the commons exists when too many individuals have the right to use a scarce resource, such as fisheries, and overuse of that resource potentially leads to its complete exhaustion. When no ownership rights to plant species exist, economic incentives stimulating individuals or countries to overexploit plant species outweigh the inducements to preserve natural habitats.[7]

Moreover, from an international law perspective, the common heritage doctrine in relation to genetic resources is not utterly convincing. Under international law, States enjoy sovereignty[8] over their lands and natural and biological resources and nothing speaks against State sovereignty over the contained genetic resources. In reality, the *problematik* relating to the nature of genetic resources in international law, started when the genetic resources contribution in commercial uses and the financial benefits of these uses were realized from the source countries. The use of genetic resources and the fact that the International Undertaking is a soft law instrument fail to shape the necessary *opinio juris* and justify customary law.

Gradually, the international challenge to the use of genetic resources and their international legal treatment started to abate. The 26$^{th}$ FAO conference amended the International Undertaking on 25 November 1991. The amended Undertaking took a hybrid position admitting that 'the concept of mankind's heritage, as applied in it, is subject to the sovereignty of the states over their plant genetic resources'. Nevertheless, developing countries still did not receive any benefits as compensation for their own efforts in preserving agricultural cultivars and natural forest areas, while industrial nations continue to affirm the rights of their companies to profit from their work in research on plants.

On 3 November 2001, the International Treaty on Plant Genetic Resources for Food and Agriculture was adopted by the FAO Conference

[6] The intellectual work on the 'commons' and 'anticommons' is relatively novel; it has been claimed that 'neither all commons nor all anticommons are tragic', but as 'extremes on a spectrum, illuminate the boundaries of private property'; see the brief and comprehensive analysis by SR Muzner, 'The Commons and the Anticommons in the Law and Theory of Property' in MP Golding and WA Edmundson (eds) *The Blackwell Guide to the philosophy of Law and Legal Theory* (Blackwell Publishing, Oxford, 2005) 148.

[7] The literature on the tragedy of the commons is vast. See, G Hardin, 'The Tragedy of the Commons' (1968) 162 Science 1244–5 (introducing the term); E Ostrom et al (eds) *The Drama of the Commons* (National Academy Press, Washington, DC, 2002); E Ostrom, *Governing the Commons: The Evolution of Institutions for Collective Action* (Cambridge University Press, 1990).

[8] See UN Charter 1945 Article 2 and UN GA Res. 2625 (XXV), Declaration on Principles of International Law Concerning Friendly Relationships and Cooperation among States in Accordance with the Charter, 1970, reprinted in 9 ILM 1292 (1970).

at its 31st Session in Rome, by Resolution 3/2001, which entered into force on 29 June 2004. This is a new legally binding instrument which seeks to ensure the conservation and sustainable management of plant genetic resources for food and agriculture, as well as the fair and equitable sharing of benefits arising from their use. The core of the International Treaty is the principle of the sovereign rights of the States over their genetic resources. Its adoption was a necessary step to revise the International Undertaking and make it compatible with the 1992 UN Convention on Biological Diversity.

### *B. Intellectual Property: International Patent Law*

Originally the concept of intellectual property rights included copyrights, authors' rights, trade marks and inventors' rights. The development of new technologies, together with the needs of modern society, substantially broadened the concept, which now includes the legal protection of broadcasting, computer programmes, biotechnology processes, and so on. Another concept raised relatively recently is the protection of genetic resources and associated traditional knowledge.

Historically, the concept of intellectual property rights was developed in industrialized societies to protect the products of human creativity, so as to provide economic incentives to those engaged in such creation. The products of such creations were recognized as property. All kinds of intellectual property rights involve the granting of a monopoly by the State to the right-holder to control the use, production and dissemination of the innovation, product, or process, protected by the right. Investments in the pharmaceutical industry for inventive, novel and useful life-saving medicines are awarded with the granting of patents. Thus, the owner of the patented medicines usually enjoys a 20-year monopoly in the market (unless his right is limited under circumstances and a compulsory license is granted).

As it stands today, patent law is the product of the legal systems of the developed countries. With the current structure of the patent system, the 'product of nature' exception does not allow a naturally occurring substance to be patented. Nevertheless, where a substance, 'previously unknown in its purified and isolated form', is refined so that the product can be distinguished in kind, and where it also demonstrates 'unexpected properties', the refined substance is patentable.

Patent law is the mechanism that presently attracts a lot of attention from benefit-sharing advocates. The Uruguay Round of the General Agreement on Tariffs and Trade (GATT) is the most recent effort to formalize an international intellectual property law system capable of addressing biodiversity prospecting, with the signing of the Trade Related Aspects of Intellectual Property Agreement (TRIPS Agreement). The subject is highly controversial. Developing countries, supported by non-governmental organizations (NGOs) and environmental groups, assert that intellectual property rights over biotechnology are a major obstacle to benefit-sharing and conservation. They argue that there is a potential to increase large multinational corporations' access to developing countries' resources, which will remain cheap and readily accessible. Industrialized countries, on the other hand, disagree; they assert that intellectual property rights protect fair rewards for innovation and increase the technological benefits of biodiversity by enhancing the commercial value of genetic resources.

The tensions are not rooted in the cultural clash between the Western and the developing societies, but in sociological and historical grounds. Industrialized countries dispose of better representational systems of ownership, because the need to protect ownership and inventiveness arose earlier. They tend to forget that their own past efforts to adapt their legal systems to new circumstances also caused conflicts and revolutions.[9] However, in their case there was no country with a more progressive legal system at the time that would try to impose its norms, as industrialized countries currently try to enforce their intellectual property standards on less developed ones.

A common criticism of the patent system is that it is highly inequitable to give patent protection only to the companies that isolate and then purify or synthesize an existing natural compound. Some argue that there is a need to lower the threshold for patentability to include products of nature, whereas others assert that it is necessary to re-conceptualize the entire idea of inventiveness.[10]

Yet how can we accommodate the interests of both developing and industrialized countries within the broader context of a political-economic debate of global proportions? If the answer to biodiversity protection is intellectual property, then there is still a long way to go. Traditional intellectual property rights do not generally protect genetic resources, because the latter do not satisfy the legal test of originality or novelty. Further, intellectual property rights are exclusive and do not allow the intellectual property right-holder to share the benefits with the provider of genetic material. The developing countries' 'biopiracy' concerns are pressing and the search for other solutions outside of patent law (contracts) continues.[11]

[9] De Soto (see previous chapter) 93.

[10] EJ Asebey and JD Kempenaar, 'Biodiversity Prospecting: Fulfilling the Mandate of the Biodiversity Convention' (1995) 28 Vanderbilt Journal of Transnational Law 711. S Prakash, 'Towards a Synergy between Biodiversity and Intellectual Property Rights' (1999) 2 The Journal of World Intellectual Property 822.

[11] On the debate see in general C Bail, 'What is the Future Role for the CBD in the Preservation of Plant Genetic Resources and Biological Diversity: A Plea for Policy Coherence' in T Cottier and PC Mavroidis (eds) *Intellectual Property—Trade, Competition and Sustainable Development* (University of Michigan Press, Ann Arbor, 2003) 515; S Biber-Klemm, 'Incentives to Bring about Conservation and Sustainable Use of Genetic Resources in the Framework of the World Trade Order' in ibid 481; SH Bragdon, 'Major Legal Regimes

The negotiations of the TRIPS Agreement, as part of the Uruguay Round, in April 1994, set minimum international standards for the protection of intellectual property. There is much discussion within the World Trade Organization (WTO) on the potentially negative effects of TRIPS on the interests of highly biodiverse developing countries. The ongoing discussions will be thoroughly examined in another chapter of this thesis.[12]

## III. TREATMENT OF GENETIC RESOURCES IN THE CBD[13]

The past and present use of biological resources have led international policymakers to perceive genetic resources as economically valuable assets, access to which must be regulated in order to avoid inefficient use. The continuing loss of biodiversity, due largely to the enormous growth of the earth's population, creates the need for appropriate management and planning of its exploitation. The role of environmental agreements in the global development process is to provide necessary regulation.

On 5 June 1992 the United Nations Conference on Environment and Development concluded the Framework Convention on Biological Diversity in Rio de Janeiro, Brazil, together with the companion Framework Convention on Climate Change. After long and acrimonious negotiations, over 150 countries, including all the major global powers except the United States, eventually signed the CBD. [14]

The CBD is the first international instrument that addresses the problem of loss of biological diversity in a global way. Until then, biodiversity issues were handled by regional conservation treaties and treaties protecting biodiversity by species or ecosystems. During the negotiations it was

Affecting Plant Genetic Resources (PGR): The Convention on Biological Diversity (CBD), the International Undertaking (IU) and the TRIPS Agreement', in ibid 441, and AK Gupta, 'Conserving Biodiversity and Rewarding Associated Knowledge and Innovation Systems: Honey Bee Perspective' in ibid 373; M Khor, *Intellectual Property, Biodiversity and Sustainable Development—Resolving the Difficult Issues* (Zed Books, London/New York, 2002); G Dutfield, *Intellectual Property Rights, Trade and Biodiversity* (IUCN Earthscan Publications Ltd, London, 2000).

[12] See Chapter 4 Section V.

[13] Previous version of this discussion was published as the article 'Analysis of the Objectives of the Convention of Biological Diversity: Their Interrelation and Implementation Guidance for Benefit-Sharing' (2006) 31 Columbia Journal of Environmental Law 243.

[14] For a short negotiating history and overview of the CBD institutions, see Secretariat of the Convention on Biological Diversity, *Handbook of the Convention on Biological Diversity* (Earthscan, London 2001) xvii–xxvi. Although most of the nations were willing to accept the treaty, at least as an exposition of general principles, the United Sates chose to dissent and risk a 'public relations disaster', because of a perception that the text worked against the US interests. See, McManis 259 and 263–264; RL Margulies, 'Protecting Biodiversity: Recognizing International Intellectual Property Rights in Plant Genetic Resources' (1993) 14 Michigan Journal of International Law 335 f.

quickly recognized that an umbrella convention combining or consolidating existing conventions would be legally or technically impossible, and so the focus was placed on a framework treaty complementing existing treaties.[15]

### *A. Key Provisions*

Scepticism exists regarding the legal effects of the CBD. Due to the hard negotiating background, the final wording of the CBD is often ambiguous. There is no doubt, by contrast, about the general philosophy of the CBD: the text of the treaty is inspired by the reciprocity theory and disclaims the common heritage of humanity model. In the preamble of the CBD, the parties affirm that 'the conservation of biological diversity is a common concern of mankind'. Countries that make efforts to conserve natural species, and countries whose industries transform raw germplasm into useful products, should be entitled to receive a portion of the profits from the sale of these products.

The objectives of the CBD are summarized in Article 1: conservation, sustainable use and the equitable sharing of benefits, part of which is appropriate access and appropriate technology transfer, reflect the spirit of mutual advantages and should be essential components of any bioprospecting effort.

Article 3 states as a general principle that States have a 'sovereign right to exploit their own resources pursuant to their own environmental policies, and a responsibility to ensure that activities within their jurisdiction or control do not cause damage to the environment of other States or of areas beyond the limits of national jurisdiction'. Repeated assertions throughout the text proclaim the sovereign rights of States over biological resources within their territory. The origin of the concept of sovereignty rights over natural resources can be found in the General Assembly Declaration on Permanent Sovereignty over Natural Resources in 1962.[16] The norms that followed the UN General Assembly Declaration of 1962 were Principle 21 of the Declaration of the Conference on the Human Environment in Stockholm in 1972, Principle 2 of the Rio Declaration on Environment and Development and finally Article 15.1[17] of the CBD.

Articles 4 and 5 specify the jurisdictional scope of the Convention and obligate parties to cooperate in the achievement of the CBD objectives. Articles 6 through 14 set forth various measures for promoting the first two CBD objectives, namely conservation and sustainable use of resources.

[15] W Lesser, *Genetic Resources—Convention on Biological Diversity: Exploring Access and Benefit-Sharing Issues* (CAB International, Oxon/New York, 1998) 3.

[16] GA Res 1803 (XVII), 14 Dec 1962.

[17] But see also the Preamble of the CBD.

Among other things, the responsibilities owed by companies to the indigenous peoples within source countries when accessing resources in these countries are addressed in Article 8 (*In Situ* Conservation). Articles 15 to 21, in turn, spell out the essential components needed to ensure the third objective under Article 1 of the CBD, in other words, the fair and equitable sharing of benefits arising out of the utilization of genetic resources.

Specifically, Article 15 requires parties to the CBD to 'facilitate access to genetic resources for environmentally sound uses'. Providing access to genetic resources is one of the primary obligations of biodiversity-rich/technology-poor parties to the CBD, whose obligations are stipulated in Articles 16 to 21.

Article 16 is the main reference to technology transfer in the CBD and it establishes the link between access to genetic resources and technology transfer to the developing world. Articles 17 to 19 obligate parties to facilitate the exchange of information, promote technical and scientific cooperation, and provide for the treatment of biotechnology and distribution of benefits. Articles 20 and 21, in turn, require parties to provide financial support and incentives for activities intended to achieve the objectives of the CBD.

The CBD is also treating intellectual property rights in plant genetic resources. According to Article 16(2), access to and transfer of technology subject to patents and other intellectual rights, 'shall be provided on terms which recognize and are consistent with the adequate and effective protection of intellectual property rights'. Further, in Article 16(5) the parties to the CBD recognize that 'patents and other intellectual property rights may have an influence on the implementation of the Convention' and commit to cooperate 'in this regard subject to national legislation and international law in order to ensure that such rights are supportive of and do not run counter to its objectives'.

Article 19 clearly announces that in exchange for access to its biodiversity, the developing world will receive a fair and equitable portion of the benefits that the North yields from their use.

To summarize, for the developed countries to continue to have access to the developing world's biological diversity, the CBD calls for active support for conservation efforts, equitable compensation to the source country and its indigenous populations, technology transfer to the source country and support for capacity-building within the source country. In this way, there is a two-way exchange: developing countries give access to genetic resources and the developed countries provide compensation and technology transfer. Each is dependant upon the other.

### *B. Analysis of the CBD Objectives (Article 1)*

The interpretation of international legal rules is a rather delicate issue, which was ultimately handled in a quite satisfactory way in Articles 31–33 of the 1969 Vienna Convention on the Law of the Treaties.[18] The Vienna Convention gives priority to literal, systematic and teleological interpretation of the treaty provisions.[19] Upon these premises, a more in depth examination of the CBD objectives is attempted in the forthcoming sections.

#### *1. Conservation of Biological Diversity*

The first objective set by the CBD is the conservation of biodiversity. Actually, the way one understands 'conservation' is critical for the understanding of the other two CBD objectives of sustainable use and fair and equitable distribution of benefits.

In everyday language, 'conservation' generally means the act or process of preservation of a material in its present state, protecting it from loss, damage or neglect. The application of this definition in the CBD, combined with its two other objectives, leads to controversial and even irrational results. If we consider that 'conservation' in the CBD conveys the meaning of preservation of the biodiversity components in their present state, the objectives of 'sustainable use' and 'fair and equitable sharing of the benefits' become obsolete. Today, 'conservation' is especially linked to the 'planned management of a natural resource to prevent exploitation, destruction, or neglect, etc', of the 'natural environment and wildlife'.[20]

The CBD itself does not give any express definition for 'conservation' and it distinguishes 'conservation' from 'sustainable use' in its objectives. But, it defines *in situ* and *ex situ* conservation: *In situ* referring to the 'conservation of ecosystems and natural habitats and the maintenance and recovery of viable populations of species in their natural surroundings, and in the case of domesticated or cultivated species, in the surroundings where they have developed their distinctive properties'; and *ex situ* meaning the 'conservation of components of biological diversity outside their natural habitats'.[21]

Accordingly, 'conservation' presents a twofold meaning in the CBD,

[18] Vienna Convention on the Law of the Treaties, (23 May 1969), 8 International Legal Materials 679; A Cassese, *International Law* (Oxford University Press, 2005) 178–180.

[19] Vienna Convention Article 31.1 reads: 'A treaty shall be interpreted in good faith in accordance with the ordinary meaning to be given to the terms of the treaty in their context and in the light of its object and purpose'.

[20] See definitions at: www.yourdictionary.com, http://www.m-w.com (Merriam-Webster OnLine Dictionary) and www.askoxford.com (Oxford English Dictionary).

[21] CBD Article 2.

depending on the protection approach, *in situ* or *ex situ*: 'conservation' can be understood narrowly to mean protection, or broadly to incorporate 'sustainable use'. Despite the textual distinction between 'conservation' and 'sustainable use', in practical terms, 'conservation' in the context of the CBD can be understood both in a narrow sense, limited to protectionist goals, and in a broader sense, involving sustainable use elements.

*In situ* conservation[22] reflecting the broad meaning of 'conservation' incorporates sustainable use and management; it encompasses efforts to preserve and manage resources valued through use in their habitat. The CBD explicitly refers to the 'establishment and management of protected areas or areas where special measures need to be taken to conserve biological diversity', the promotion of 'environmentally sound and sustainable development', protection of indigenous management practices, amongst others.[23] In sum, *in situ* conservation presupposes human intervention and sustainable management of the resources, and the advancement of this strategy is vital for conservation purposes.

Sustainability considerations dictate that use of resources is very often necessary for the better preservation of biodiversity. After all, human beings constitute part of the earth's biodiversity, which has a certain dynamic and there are diverse evolutionary processes happening within the system. Human intervention is sometimes necessary for the conservation of this diversity. Domestication and breeding of animals, farming and agriculture, although they transform the original system, still promote its functioning under the precondition of sustainable use. Unsustainability problems arise when people misuse or abuse their surrounding resources.

Conversely, *ex situ* conservation stands for the narrow understanding of 'conservation', because it involves the preservation of the biological material in its present state in gene-banks.[24] Taking into account the dynamic of eco-systems and resources, but also the degeneration and extinction of many species, the conservation in 'museum-like' perfection of some resources in banks and collections was deemed necessary for scientific and food-safety reasons.

*In situ* conservation methods comprising human intervention and management, aims at creating conservation by use of economic means. *Ex situ* conservation does not imply either great economic incentives or equitable sharing of benefits generated by the existing gene-banks world wide with the provider countries, because most *ex situ* collections acquired their genetic resources prior to the entry into force of the CBD, and, hence, remain out of the scope of the CBD and its stipulations.

22 CBD Article 8.
23 ibid *lit* a, e, and j respectively.
24 CBD Article 9.

Consequently, 'conservation' is not just understood as an attempt to protect any particular *status quo* in perfection, like in gene-banks; but rather to conserve and enhance the ability of ecosystems to develop and regenerate themselves as living systems. The 'sustainable use' objective of the CBD complements and supports the notion of conservation.

### 2. *Sustainable use of biodiversity*

The terms 'sustainability', 'sustainable development', 'sustainable use', etc, are to be found in a vast array of international instruments dealing with resource use and development. They are present in almost every discourse dealing with environmental and developmental concerns. The CBD is not an exception and the sustainable use of the components of biological diversity is declared in Article 1 as one of the principal objectives of the convention. At present, the 'sustainability' discourse has replaced most of the previously only 'environmental' discourse, because of the acknowledgement that the 'environment' cannot be considered in isolation of other developmental, economic and social concerns, but rather as a cross-cutting issue in human activities.

#### a) Sustainability

The term sustainability can be generally defined as the 'capacity of being borne or endured, upheld or defended, the capacity that something has to be sustained into the future'.[25] In the past, the term was mainly applied in relation to productivity in different contexts, such as development, society, industry, agriculture, cities, etc. Today, the 'sustainability' dimensions have surpassed the limits of use in the framework of productivity; 'sustainability' appears frequently in discourse within different conceptual frameworks and cultural practices, political, economic, social, technological and environmental.[26] It also intersects with applied ethics issues, such as environmental ethics, bioethics, international justice and equity, and war and peace.

The increased use of the term can be associated to the significant increase of natural resources use, two hundred generations ago, and the immense acceleration of this use due to the industrialization and urbanization in the 20th century.[27] Ruckelshaus[28] makes the distinction between 'aboriginal sustainability', 'transitional unsustainability' and 'consciousness of advanced sustainability'.

[25] See The Oxford English Dictionary, Second Edition, Clarendon Press, Oxford 1989; N Dower, 'Development Issues, Environmental' in RF Chadwick (ed) Encyclopedia of Applied Ethics (Academic Press, San Diego, 1998) 767.

[26] SR Carpenter, 'Sustainability' in Chadwick (ed) (n 25) 279 f.

[27] ibid 277.

[28] WD Ruckelshaus (1989) 'Toward a Sustainable World', Scientific American 166–175, cited by Carpenter in Chadwick (ed) (n 25) 279 and 281.

Human beings were absent in most of the period of earth life. The appearance of humans and the birth of human culture led to the increase of 'basic biometabolism of life' through human technology ('technometabolism'); humans lived with the aid of their tools in a state of 'aboriginal sustainability'. During this era, which is the longest and lasted until the 20th century, humans interfered with nature and introduced modifications (such as agriculture and animal breeding).

The period of 'transitional unsustainability' commenced in the 20th century, and featured the emergence of cities, massive increases in population, and a sudden expansion of industrialization and energy. As population growth and resource use do not occur in a uniform way, international unsustainability problems appeared. The recognition of these problems, and the acute need for international cooperation and action, introduces the postindustrial era informed by 'the consciousness of advanced sustainability'. This era will be established when the notion of sustainable development acquires a complete and wide-ranging treatment integrating the role of values in human activities.

Based on this historical line of reasoning, Carpenter argues that sustainability of human populations involves 'the persistence through time of the diversity of human communities and ethical ideals of human flourishing, the dynamically balanced development of economic enterprise, and the preservation and regeneration of ecological systems and resources that sustain that development'.[29] Apparently, the notion of 'sustainability' is not simple to define, but it encompasses an array of interrelated facts and processes. It is a word with a leading role in the present international political discussions over developmental, environmental and economical issues. 'Sustainability' stands for the connection of humans to nature, for the relation of different generations in association with the use of natural resources, and in general it is a part of a whole modern ideology developed in the end of the 20th century.

### b) Sustainability and international law initiatives

Sustainability was first enunciated, as described below, in the Brundtland Report, and later developed by consecutive Resolutions and Declarations. Given the difficulties in defining the exact criteria of sustainability, the concept of sustainable use is closely related to achieving policy integration between environmental and economic benefits. More specifically, in international law the origins of 'sustainability' lead back to forestry and fisheries conservation regulatory efforts; 'sustainability' stood for the maintenance of the long-term productivity (*Nachhaltigkeit* in German, *durabilité* in French) of these resources.

[29] ibid 276.

Precursors of 'sustainable development' can be traced in the 1971 Ramsar Convention and the 1972 Stockholm Conference. Under Article 3.1 of the 1971 Ramsar Convention,[30] contracting parties agree to 'formulate and implement their planning so as to promote the conservation of the wetlands included in the List, and as far as possible the wise use of wetlands in their territory'. The term 'wise use' was pioneering at the time of drafting of the Ramsar Convention, and its application is actually essential for the maintenance of biological diversity and human well-being. Later, 'wise use' was expressly associated with conservation and human use. The Conference of the Parties at its 3rd meeting in Regina, Canada, 1987, adopted the following definition: 'the wise use of wetlands is their sustainable utilization for the benefit of mankind in a way compatible with the maintenance of the natural resources of the ecosystem'. At the same time, 'sustainable utilization' was defined as: 'human use of a wetland so that it may yield the greatest continuous benefit to present generations while maintaining its potential to meet the needs and aspirations of future generations'.[31]

At the Declaration of the 1972 United Nations Conference on the Human Environment, held in Stockholm,[32] one encounters claims of entitlement for 'freedom, equality and adequate conditions of life, in an environment of a quality that permits a life of dignity and well-being', while one 'bears a solemn responsibility to protect and improve the environment for present and future generations'.[33] Also, developed countries are encouraged to provide for 'financial and technological assistance as a supplement to the domestic effort of the developing countries'.[34]

The above propositions were emphasized by developing countries, and in order to counterbalance this set of obligations, developed countries insisted on reaffirming the sovereignty principle:

> '[T]he states have the sovereign right to exploit their own resources pursuant to their own environmental policies, and the responsibility to ensure that activities within their jurisdiction or control do not cause damage to the environment of other States or of areas beyond the limits of national jurisdiction'.[35]

[30] The 1971 Convention on Wetlands of International Importance especially as Waterfowl Habitat, Ramsar, Iran 2 February 1971, as amended by the Protocol of 3 December 1982 and the Amendments of 28 May 1987, at http://www.ramsar.org/key_conv_e.htm (13/09/08).

[31] The Ramsar concept of 'wise use', Ramsar Convention, The Ramsar Info Pack, Ramsar Information Paper No 7.

[32] Declaration of the United Nations Conference on the Human Environment, Stockholm 16 June 1972.

[33] ibid Principle 1.

[34] ibid Principle 9.

[35] ibid Principle 21.

This discord in the Stockholm Declaration reflects the classical dichotomy between positive and negative rights echoing the disparate claims of developing and developed countries. The same dichotomy appears in the CBD as well, where for example, developed countries commit to transfer technology to developing countries, while intellectual property rights ought not to be affected (Article 16). This is actually a conventional phenomenon in international legal texts, where diverging interests are to be accommodated in a single text, and in the end the text seems vague and contrasting in content, its interpretation, implementation and enforcement become complex issues, and it loses credibility.

It was not until the United Nations General Assembly Process of preparation of the 'Environmental Perspective to the Year 2000 and Beyond',[36] that 'sustainability' was explicitly employed. The special Commission on Environment and Development had the mandate to 'propose long-term environmental strategies for achieving sustainable development'[37] and to examine 'ways in which concern for the environment may be translated into greater cooperation among developing countries and between countries at different stages of economic and social development'.[38] The resulting report of the process undertaken by the World Commission on Environment and Development, known under the title 'Our Common Future' or Brundtland Report[39] gave impetus to the idea of 'sustainable development' defining it as 'meeting the needs of the present without compromising the ability of the future generations to meet their own needs'.[40] The report proposes policies in six areas or 'common challenges'; population, food security, species and ecosystems, energy, industry and urbanism. Global environmental issues, like desertification, deforestation, acidification, global warming, and ozone layer depletion are 'failures of management'. The interests clash between developing and developed countries are again portrayed in the Report, which raised a lot of criticism and was characterized as a 'commendable and constructive compromise'[41] between the conflicting agendas.

'Sustainability' was in reality established with the World Conference on Environment and Development in 1992. The Rio Declaration places human beings at the centre of concerns for sustainable development and defends their entitlement to a healthy and productive life in harmony with nature.[42]

36 Process of preparation of the Environmental Perspective to the Year 2000 and Beyond, GA Res. A/38/161 of 19 December 1983.

37 ibid Article 8a.

38 ibid Article 10.

39 World Commission on Environment and Development, 'Our Common Future' (University Press, Oxford, New York, 1987), named Brudtland Report after its Chair, Gro Harlem Brundtland, prime minister of Norway.

40 ibid 43.

41 Carpenter in Chadwick (ed) (n 25) 279.

42 Rio Declaration, United Nations Conference on Environment and Development (Earth Summit), Rio de Janeiro (Brazil), 3–14 June 1992, Principle 1.

Obviously, by highlighting sustainable development and placing human beings in a central place, the Declaration embraces the 'anthropocentric approach' for the protection of the environment. There is an ideological, philosophical and theological conflict between anthropocentric and ecocentric justification for the protection of the environment, but ultimately this conflict is merely a pseudo problem. The anthropocentric focuses on humans, whereas the ecocentric advocates the protection of the environment as an intrinsic value.[43] The subject of protection in the various domestic and/or international instruments reveals the anthropocentric or ecocentric choice of the domestic and/or international policymaker, while sometimes, both ideologies are present.[44] However, even if sometimes a policy choice appears to be based on the altruistic ecocentric reasoning, it can also be very well justified under the anthropocentric view. If for instance the objective of the law is the 'conservation of biological diversity', there seems to be an ecocentric underpinning; however, thinking that the 'conservation of biological diversity' is a human need or a gene reserve serving humans, the anthropocentric view is a legitimate basis as well. In the end, it is very difficult to imagine that human beings in their cognitive capacity can retreat from the anthropocentric standpoint and think really in an ecocentric manner. Because of the complexity of nature and the fact that humans belong to it, the differences between the two angles ultimately disappear.[45]

At the 2002 World Summit of Sustainable Development (WSSD), the Johannesburg Declaration corroborates the commitment to sustainable development[46] as the obligation to 'build a human, equitable and caring global society cognizant of the need for human dignity for all'.[47]

### c) Sustainability, sustainable use and sustainable development

At present the terms sustainability, sustainable development and sustainable use are used interchangeably. Sustainability is to some extent the generic term, which encompasses the sustainable use and the sustainable development. Sustainable use of natural resources is one of the foundations for sustainable development. Sustainable development is more oriented to the evolution of human populations, the persistence through time of the diversity of human communities and ethical ideals of human flourishing, the dynamically balanced development of economic enterprise and the preservation and

[43] M Klöpfer, *Umweltrecht* (CH Beck, München, 1998) 19.

[44] eg in the German Federal Law for the Protection against Emissions (Immissionsschutzgesetz–ImSchG) subject of protection according to Article 1 is 'humans and also animals, plants and other matters' (translation by the author).

[45] Klöpfer (n 43) 20.

[46] Johannesburg Declaration on Sustainable Development, para 1.

[47] ibid para 2.

regeneration of ecological systems and resources that sustain that development. Because of its spatial and temporal dimensions, 'sustainability' is a geometrical concept. It comprises different dimensions of the interrelationship between humanity and nature, and between humans of the same generation in different places on earth suggesting the ecological, economical and social elements of this relationship; furthermore, time—past, future and present—is an inherent aspect of 'sustainability', as future generations should be considered in current decision-making.

Every generation exists in relationship to the previous and the following generation, and hence 'we must develop a planetary ethos which recognizes that we share the planet with all communities throughout time'.[48] 'Sustainability' implies the common responsibility and the necessity of *intra*generational and *inter*generational equity.[49] At this point, we can pin down the conjuncture of sustainability with equity and justice concerns,[50] which entail that 'sustainability' goes beyond ecological, economic and social concerns to a moral valuation.

'Global thinking' as awareness-raising on the interdependence of natural resources, the dependence of human beings on natural resources, and the global effects on humanity, is directly implied by present sustainability concerns, and calls for 'global action', for cooperation and coordination in the management of environmental and developmental issues. At the same time, local action is necessary for the effective fleshing out of the global guidelines.[51]

The combination of sustainability with development first shows up at the IUCN World Conservation Strategy in 1980,[52] but its use was established,

[48] E Brown Weiss, *In Fairness to Future Generations: International Law, Common Patrimony, and Intergenerational Equity* (United Nations University and Transnational Publishers Inc, 1989) 291.

[49] ibid 21; *Inter*generational equity refers to relationships between generations, whereas *intra*generational refers to relationships among members of a generation. The theory of intergenerational equity should not be confined only to the sets of relationships between the past and the present on the one hand, and the present and the future generations on the other, but should also encompass 'a parallel set of planetary obligations and planetary rights that are *intragenerational*'.

[50] Carpenter in Chadwick (ed) (n 25) 287; Brown Weiss (n 48) 39; Brundtland Report (n 39) 43.

[51] 'Think Globally, Act Locally' or 'Local Action moves the World' were the motto at the WSSD in Johannesburg 2002, where the states underlined the empowerment of local administration in sustainability questions and local authorities held the Local Government Session (LGS) as part of the whole summit; RRM Verhick, 'Why the Global Environment needs Local Government: Lessons from the Johannesburg Summit' (2003) 35 Urban Lawyer 471; CR McManis, 'Intellectual Property, Genetic Resources and Traditional Knowledge Protection: Thinking Globally, Acting Locally' (2003) 11 Cardozo Journal of International and Comparative Law 547.

[52] IUCN/UNEP/WWF, *World Conservation Strategy: Living Resource Conservation for Sustainable Development*, prepared by the International Union for Conservation of Nature and Natural Resources (IUCN), with the advice, cooperation and financial assistance of the

as previously seen, with the Brundtland Report and then at the World Conference for Environment and Development.

There is not a common understanding for the formula of 'sustainable development'. Each discipline lends a different meaning, whereas some might stress 'sustainable' and others 'development'. The definition of development alone is not a simple exercise. A conventional definition of 'development' as 'growth' renders 'sustainable growth' an oxymoron, because the concept of 'growth' knows no limits and the concept of 'sustainable' implies certain limits. Daly makes the point that 'as long as our race inhabits a finite planet, and a very thin biospheric shell of it, both unlimited economic and population growth are unsustainable'.[53]

The Brundtland Report itself puts policy emphasis on 'growth'. However, 'sustainability' and 'sustainable development' were still in embryonic use; several 'sustainability' aspects were not sufficiently studied and integrated in the environmental discourse in 1987. And last, but not least, industrialized countries' and industry's agendas were at that time far more influential than at present. Despite the general 'economic growth' orientation of the Report, there appears a tendency for 'elevation of sustainable development to a global ethic'.[54] Indeed, development is primarily and centrally a process of economic growth,[55] but behind this dominant 'growth' conception, we can discover a process of change in society from a worse to a better situation.

The elements of sustainable development broadly traced in Agenda 21 are conservation of natural resources and of the environment, economic growth, equity in the allocation of resources, participatory and more democratic political systems, population levels in harmony with the productive potential ecosystems, and adoption of life-styles within the planet's ecological means.[56] Accordingly, beyond 'growth', the 'improvement' facet of

United Nations Environment Programme (UNEP) and the World Wildlife Fund (WWF), and in collaboration with the Food and Agriculture Organization of the United Nations (FAO) and the United Nations Educational, Scientific and Cultural Organization (UNESCO), IUCN (IUCN/UNEP/WWF, Gland, 1980).

[53] Daly (1991), 'Operational Principles for sustainable development' *Earth Ethics* (Summer), cited by Carpenter in Chadwick (n 25) 279.

[54] Brundtland Report (n 39) 308.

[55] In accordance with the principal definitions of 'development' in the dictionaries, as 'growth', 'unfolding' and 'evolution'. See *The Oxford English Dictionary* (Second Edition, Clarendon Press, Oxford, 1989); the same in French for *développement* (*croissance, déroulement, évolution*) and German for *Entwicklung* (*Wachstum, allmählich entstehen, auf ein höheres Niveau heben*), see Le Robert, *Dictionnaire de la Langue Française* (Deuxième édition, Paris 1985) and Duden, *Das große Wörterbuch der deutschen Sprache* (Mannheim, 1976) respectively.

[56] Agenda 21, Report of the United Nations Conference on Environment and Development (Earth Summit), Rio de Janeiro (Brazil), 3–14 June 1992, UN Doc A/CONF.151/26.

development gives rise to a series of alternative ethical considerations,[57] to development ethics and to environmental ethics.[58] Environmental ethics is an active area of inquiry evolved in the last 20 years. It is a response to the recognition that human activities are affecting and often disturbing the physical environment and that therefore there is a need to collectively change human practices. It focuses on the natural world as a source of value in itself and also on our responsibilities to future generations.[59] The pursuit of development and the protection of the environment are often seen to come into conflict with each other.

'Sustainable development' is a serious attempt to describe and integrate both sets of concerns: growth and ethical considerations. There is a desire to define 'development' beyond 'growth' and economic usage in a holistic way of overall quality of life. However, with the exception of a few optimists, commentators are not very satisfied with the evolution of the term 'development' arguing that during these last 30 years, the normative accounts of 'development' do not escape the economic growth paradigm. Criticizing the economic growth understanding of 'sustainability', Kothari identifies four primary criteria for sustainable development:

> ... a holistic view of development; equity based on the autonomy and self-reliance of diverse entities instead of a structure of dependence founded on aid and transfer of technology with a view to 'catching up'; an emphasis on participation; and an accent on the importance of local conditions and the value of diversity.[60]

Similarly, Engel, who supports Kothari's insistence that sustainable development is an issue for applied ethics, proposes a comprehensive social paradigm, consisting of five operational principles; integration of conservation and development, satisfaction of basic human needs, achievement of equity and social justice, provision for social self-determination and cultural diversity, and maintenance of ecological integrity.[61]

[57] Economic growth in a society needs to obey to some basic ethical standards to promote the interests of all citizens and reduce poverty standards (*good* life for *all*). The culminating function of ethics in our century gave rise to the different theories in the framework of various disciplines, like J Rawls, *A Theory of Justice* (The Belknap Press of Harvard University Press, Cambridge, Massachusetts, 1971, rev ed 1999) in political philosophy, or the principle of equity in economics, see further down in this chapter.

[58] See JR Engel, 'Introduction: The ethics of sustainable development' in JR Engel and JB Engel (eds) (1990) *Ethics of Environment and Development—Global Challenge and International response* (Belhaven Press, London 1990) 1.

[59] See also the theory of intergenerational equity, Brown Weiss (n 48).

[60] R Kothari, 'Environment, Technology and Ethics' in Engel and Engel (eds) (n 58) 34.

[61] Engel in Engel and Engel (eds) (n 58) 8–9.

A notable activity in the field of environment and ethics during the last decade is the conjunction of 'sustainability' with ethics.[62] Among various activities, it is worth mentioning the Food and Agriculture Organization, which recognizing the variety of ethical questions of relevance to food security and sustainable rural development, has designated 'Ethics in Food and Agriculture' as a Priority Area for Interdisciplinary Action, and established an internal Committee on Ethics in Food and Agriculture to guide the Organization's actions in this regard. The Director-General has also established, under Article VI.4 of the FAO Constitution, an independent Panel of Eminent Experts on Ethics in Food and Agriculture, to advise him on pressing issues of ethics in food and agriculture.[63]

d) Sustainability and biodiversity concerns

Specifically in the context of genetic resources, both the CBD and the FAO International Treaty refer to 'sustainable use' of the components of biodiversity[64] and of plant genetic resources for food and agriculture[65] respectively.

The CBD definition of its objective of 'sustainable use' builds upon the Brundtland Report pattern, as 'the use of components of biological diversity in a way and at a rate that does not lead to the long-term decline of biological diversity, thereby maintaining its potential to meet the needs and aspirations of present and future generations'.[66] Furthermore, the CBD expressly relates conservation to sustainable use in several of its provisions: it calls for their integration 'into relevant sectoral and cross-sectoral plans, programmes and policies'[67] and into national decision-making;[68] or it encourages the contracting parties to 'endeavour to provide the conditions needed for compatibility between present uses and conservation of biological diversity and the sustainable use of its components'[69] and to 'respect, preserve and maintain knowledge, innovations and practices of indigenous and local communities embodying traditional lifestyles relevant for the

[62] See for instance the International Development Ethics Association—IDEA (http://www.development-ethics.org/) and the International Union for the Conservation of Nature—IUCN Ethics Working Group, which in 1996 issued the publication 'Earth Ethics', which, according to IUCN, 'promotes alternative models for sustainable communities and lifestyles, based in ecospiritual practices and principles [to solve] the problems that face the world today humanity must undergo a radical change in its attitudes, values, and behavior [...] In response [...], a new global ethics is taking from, and it is finding expression in international law', see MS Coffmann, 'The IUCN: From the UN to Your Back Yard' *Sovereignty International* (1998) published on the Internet at http://www.sovereignty.net/p/ngo/iucn.htm (29/12/2008).

[63] Food and Agriculture Organization of the United Nations, Ethics in Agriculture, at: http://www.fao.org/ethics/ (13/09/08).

[64] CBD Article 1.

[65] ITPGRFA Article 1.1.

[66] CBD Article 2.

[67] CBD Article 6.

[68] CBD Article 10.

[69] CBD Article 8i.

conservation and sustainable use of biological diversity'.[70] Finally, the CBD takes into account the economic aspects of the sustainable use in the incentive measures[71] and the access and benefit-sharing.[72]

The FAO International Treaty, dealing specifically with agro-biodiversity, lacks a definition of its 'sustainable use' objective, but enumerates an indicative list of policy and legal measures for the achievement of this goal[73] It explains that sustainable use of plant genetic resources for food and agriculture may include such measures as the promotion of diverse farming systems, of plant breeding efforts adapted to local conditions with the participation of farmers, and of the use of local and locally adapted crops.

Plainly, conservation concerns with relation to biodiversity are related to sustainability concerns, use and management.

### 3. *Fair and Equitable Sharing*

After analyzing the first two objectives of the CBD and establishing a clear link between conservation and sustainable use, the difficult task of investigating the third objective of 'fair and equitable' benefit-sharing comes along. It is an innovative element and a *leitmotiv* in the CBD inspiring its provisions of an economic nature. However, the CBD does not specify how its 'fairness and equity' objectives are to be interpreted and implemented; the guidelines for the implementation of its third objective are very broad, leaving a large part of its application upon the discretion of the domestic policy-maker.

In dictionaries of the English language, justice, equity, equitableness and fairness have very similar definitions to the qualities of being just, equitable, fair, impartial and unbiased;[74] they are related with standards of what is reasonable, right and proper. The similarities these terms exhibit in their description in dictionaries and their common understanding signals the affinity of the notions and their interrelation. However, the dictionary definitions remain circular and the sense of the three terms is not identical, even though they are used alternatively by some authors.[75] Their precise content remains ambiguous, but it will be shown here that 'fairness' and 'equity' in the CBD represent the practical aspects of justice, in terms of process and outcome of distribution.

[70] CBD Article 8j.

[71] CBD Article 11.

[72] CBD Article 15.

[73] ITPGRFA Article 6.

[74] See *The Oxford English Dictionary* (Second Edition, Clarendon Press, Oxford, 1989). Also, http://www.yourdictionary.com/.

[75] See eg Brown Weiss (n 48).

a) Equity in disciplines other than law

From the three concepts of justice, fairness and equity, the latter is the one we come across in most other disciplines other than political philosophy and law. Depending on the setting, equity has a different meaning. Philosophers, moralists and sociologists use the term in different ways to economists, politicians and diplomats. There is actually an important literature in economics that addresses normative issues in social and economic justice. A variety of economic theories and approaches, such as the theory of inequality and poverty measurement, welfare economics, the theory of social choice, the theory of bargaining and of cooperative games, and the theory of fair allocation, provide many insights in matters of equity.[76]

More specifically, in welfare economics equity is used in relation to efficiency. According to economic theory, market or economic efficiency, measured as the difference between benefits and costs, is achieved by an ideal market situation with perfect competition, which leads to the optimal situation, known as Pareto efficiency. In a Pareto optimal situation, a competitive market will guarantee the efficient allocation of resources among consumers and efficient usage of factors of production, and nobody's situation can be improved without worsening the situation of somebody else. However, this optimality is not related to distributional issues; a situation is Pareto-efficient, not only if there is an equal distribution among negotiating parties, but also if one party gets all and the other nothing. The additional criterion of equity is introduced here as the fundamental moral principle underpinning economic theory to correct this morally undesired situation. Equity inserts an element of distributive justice as a corrective complement of the pure market efficiency, but equity depends on a number of variables, which render the determination of an equitable result a difficult task.[77]

In sociology, the concept of equity has served to justify the legitimization of authority in bureaucracies. Weber (1947), Simon (1957) and Presthus (1960) have regarded equity as the moral foundation for the premise that

[76] M Fleurbaey, 'Economics and Economic Justice' *The Stanford Encyclopedia of Philosophy* at http://plato.stanford.edu (article first published Fri 28 May 28 2004; substantive revision Mon 9 June, 20).

[77] RN Stavins, 'Environmental Protection and Economic Well-Being: How does (and how should) government balance these two important values?' Presentation at: *Tackling the Critical Conundrum: How do Business, Government and Media balance the Competing Values of Economic Growth and Healthy Environment?* The Aspen Institute, Program of Energy, the Environment, and the Economy, Aspen, Colorado, 10–12 Janaury 2004; E Fehr and KM Schmidt, 'A Theory of Fairness, Competition, and Cooperation' (1999) 114(3) The Quarterly Journal of Economics, 817; Rawls (n 57) 57 f, also adopts the concept of efficiency that is associated with the Pareto efficiency in the field of economics.

those in authority have the right to demand obedience and that those subject to authority, the obligation to obey. Further research took place to define the criteria societies use for the distribution of authority.[78]

In social psychology, equity theory is a method to understand the concept of distributive justice held by a specific society. It was only in the last 50 years that the concept of justice emerged in social psychological theory and research, with Homans (1961), Blau (1964) and Adams (1965).[79] The work of the latter gave impetus for a large volume of research in the 1960s and 1970s that led to the development of a version of equity theory by Walster and her colleagues in 1973. Social psychologists study our perceptions of justice and injustice and the human emotional responses they cause. Their research suggests that the human perception of injustice is one of the most powerful motivators in the human psyche. People make some investments or inputs in justice, for which they expect to receive some rewards or outputs.[80] The rule of equity is strongly related to distributive justice and it is one of the several specific rules that may be employed to determine what is a fair distribution of outcomes. In fact, equity is termed as the proportional equality principle as conceived by Aristotle, and it constitutes the most fundamental principle of justice. In other words, the rule of equity claims that one's outcome should be proportional to one's input or contribution. The study of equity is essential for the understanding of the various types of human relationships and organizations, whatever form these have, ie legal, political, economic, religious, social, or cultural.[81]

b) Justice: from morality to social structure and law

Although, it is beyond the scope of this thesis to go deep into philosophical questions, it is necessary to explore some issues related to moral philosophy and political thought to explore the meaning of 'fair and equitable' in legal studies. In this regard, 'justice' presents many definitional difficulties and the everlasting search for the real meaning of the concept starts with the ancient Greeks.[82]

[78] CG McClintock and LJ Keil, 'Equity and Social Exchange' in J Greenberg and RL Cohen (eds) *Equity and Justice in Social Behavior* (Academic Press, New York, 1982) 337.

[79] J Greenberg and RL Cohen, 'The Justice Concept in Social Psychology' in ibid, 1 f.

[80] FJ Garcia, 'Trade Justice and Security' Boston College Law School Research Paper No 84, 28 October 2005.

[81] McClintock and Keil in Greenberg and Cohen (eds) (n 78) 1 f and 337 f.

[82] TD Campbell, *Justice* (Macmillan, Basingstoke, 2001); SC Kolm, *Modern Theories of Justice* (MIT Press, Cambridge, Mass, 1996); J Finnis, *Natural Law and Natural Rights* (Oxford University Press, 1980) 161f; BA Mellers, 'Introductory Remarks' in BA Mellers and J Baron (eds) *Psychological Perspectives on Justice: Theory and Applications* (Cambridge University Press, 1993) 1.

There is a certain and intimate link between justice as a human quality, and as a political organization and law,[83] but this link is not very evident.[84] In morality and ethics[85] justice comes as a human quality of the individual. However, man being Ον πολιτικόν (*on politikon*, ie political being), his character defines his behaviour to the others and to the community. The virtue of justice is reflected in the attitude of the individuals within the sociopolitical organization and the efforts for the achievement of the common good.[86]

The natural law theory according to the Aristotelian/Thomist tradition provides a comprehensive analysis for the concept of justice and the association of law to morality.[87] For Aristotle, justice is a virtue that affects the relations between nature and free equals in political community.[88] The natural law theory implies that there are certain principles of true morality, and certain laws of nature, which for Aquinas are 'eternal'.[89] Some of these laws have been discovered by human reason and articulated for all, and some have not been codified. In any case, manmade law that conflicts with natural law is not valid law. Apart from ancient Greek philosophers, like Aristotle and Plato,[90] this concept of the laws of nature

[83] It should be here noted that in Greek justice (dikaio/dikaiosyni—Δικαιοσύνη) and law (dikaio –Δίκαιο) is the same word.

[84] HL Hart, *The Concept of Law* (Clarendon Press, Oxford, 1972) 151–180, for a brief outline intending to show the distinction within the general sphere of morality of the specific idea of justice and the special features which account for its connection with law; RA Newman, 'Introduction' in RA Newman (ed) *Equity in The World's Legal Systems: A Comparative Study Dedicated to René Cassin* (Établissements Émile Bruylant, Brussels, 1973) 15–29; EU Petersmann, 'Theories of Justice, Human Rights and the Constitution of International Markets—Symposium: The Emerging Transnational Constitution' (2003) 37 Loyola of Los Angeles Law Review 407. On the various efforts to explain the concept (Justice as entitlement, as rights, as contract, as efficiency, etc), see Campbell (n 82) and Kolm (n 82).

[85] The distinction between morality and ethics is by no means a conventional matter: ethics and morality are often used interchangeably and they are going to be used interchangeably in this text. However, there is a need to clarify that 'morality' most properly refers to judgments and actions regarding what is right or good, and 'ethics' to the reasoning such judgments and actions require, see WK Frankena, *Ethics* (Englewood Cliffs, New Jersey, 1973) 3, cited by Engel in Engel and Engel (eds) (1990) 2.

[86] See Finnis (1980) 163–165, who also understands justice as a quality of character, as a practical willingness to favor and foster the common good in one's communities.

[87] The reason we usually speak o Aristotelian/Thomist tradition is because Thomas Aquinas (1224/25-1274) was one of the most interesting commentators of Aristotle's ('the Philosopher' is his words) greatest works and his philosophy has inspired many after his death; see Aquinas, *Political Writings* (edited by RW Dyson) (Cambridge University Press, 2002) xxiii f; JP Dougherty, 'Thomism' in Chadwick (n 25) 365.

[88] Aquinas ibid 168 f.

[89] ibid xxxii.

[90] Plato, *The Republic* (edited by GR Ferrari) (Cambridge University Press, Cambridge, 2000); for Plato, justice is an order of 'health, beauty and vigour in the soul' (Book 4, 444e) and is actually the cause of existence of the other three cardinal virtues: Self-Discipline, Wisdom, and Courage (Book 4, 427e).

leads back to ancient Greek poets[91] and can be later found at Roman stoics.[92]

Hence, social justice goes beyond the primary understanding of justice as a moral human quality; it hearkens to rationality and is detached from emotional and spontaneous actions. Social justice maintains or restores a balance, equilibrium or proportionality among the individuals who participate in society, in accordance with generally acceptable rules in time and place, social good or welfare. Justice entails a sort of obligation and it comprises the element of equality, meaning equal treatment (*Gleichbehandlung*).[93] Whether true or not, justice is at least a necessary condition to be met by any legislative choice which purports to serve the 'common good'.[94] Social changes or laws do not promote the welfare of all individuals alike, but certain choices are made among competing claims of different social classes and interests, in order to make the choice that better serves the common good. However, there is a morally accepted status quo, which should remain in balance between two or more individuals and any wrong action should be restored. In the name of social good or welfare the 'just' law may enforce compensation from one who has injured another, even when morally as a matter of justice, compensation might not be thought due.[95] Forms of general moral criticism, like 'wrong', 'bad', or 'wicked' differ from the characterization 'unjust'; a 'wrong' or 'bad' law is not necessarily 'unjust', but an 'unjust' law is a 'bad' law.[96]

c) Domestic justice: fairness and equity in law deriving from justice

Equity appears in different areas of law, without a clear explanation of its meaning. In public administrative law, the state has to function in a fair and equitable manner towards the citizens, providing for fair and equitable opportunities. In contract law, we encounter equity and fairness in agree-

[91] In ancient Greek tragedy, the characters very often obey to this 'higher law' of morals and act contrary to the commands of legitimate power. For example, in Sophocles' 'Antigone', Antigone refuses to obey Creon's commands, because these are incompatible with her conscience and religion, and they are not 'right'.

[92] Like in Seneca, see GM Galles, 'Cicero on Justice, Law, and Liberty' A Journal for Western Man, Issue XXX, 26 January 2005.

[93] Hart (n 84) 155 f lays emphasis on the motto 'treat like cases alike and different cases differently'.

[94] ibid 162, 163.

[95] Strict liability rules, where liability is independent of the intention to injure or failure to take care, are defended on the ground that it is in the general interest of the society that those accidentally injured be compensated; and the easy way to achieve such compensation is to place the burden on those whose activities, however carefully controlled, resulted in the accident.

[96] Hart (n 84) 153 f gives the illustrative example of the apartheid regimes which forbade by law coloured people to use the public transport or the parks; that was not simply a 'bad' law, but it qualified for an 'unjust' law, because it discriminated between persons who are in all the relevant respects alike.

ments, where one party needs special protection (consumer or employee) or in relation to the fair and equitable balance of rights and obligations.

The origins of equity go back to the Aristotelian Επιείκεια (*Epieikeia*) and later to the Roman *aequitas*, but it is an old and familiar legal concept in Chinese, Hindu, Judaic and Islamic law and philosophy, mainly associated with what is moral and just.[97] Despite all these major legal traditions, it appears that equity in current international practice evolved from the Western legal system, and principally from the common law practice. The Western experience begins with Aristotle's doctrinal legacy, which then affected profoundly the Roman and Common law traditions.[98] Aristotle distinguishes between universal and particular justice.[99] Universal justice stands for the lawful and the complete virtue in relation to our neighbor; particular justice stands for the fair and equal. Universal justice relates to honour or money or safety in general; particular justice (or equity or fairness) is concerned with all the objects which the good man is concerned with. One could conclude that equity and fairness are the practical facets of the virtue justice, in other words the universal justice.

Aristotle further divides particular justice (or the fair and equal) into corrective and distributive justice.[100] Corrective (or remedial or commutative) justice is further reduced to voluntary (lending, selling, etc) and involuntary (force, theft, etc) transactions. The application of corrective justice is assigned to the judiciary; judges are expected to re-establish the balance in transactions with arithmetical progression, namely they have to reinstate the middle point of equality in the private, voluntary and involuntary, exchanges.[101] Distributive justice is applied to the allotment of honor, wealth, and other goods.[102] Hence, according to Aristotelian doctrine, fairness and equity emanate from justice and both stand for its practical application in terms of compensation and distribution.

In its corrective form, Aristotelian equity is defined as the corrective law *in casu*, permitting the judge to exempt certain cases from the strict application of the law, in order to achieve just result.[103] In this corrective sense

[97] Newman in Newman (ed) (n 84) 27.

[98] CR Rossi, *Equity and International Law: a Legal Realist Approach to International Decision-making* (Transnational Publishers, Inc, Irvington, New York, 1998) 22.

[99] Aristotle, *The Nicomachean Ethics* (edited by Sir D Ross) (Oxford University Press, 1959) Book V, Chapter 1, 106 f.

[100] ibid Chapter 2.

[101] ibid Chapter 4.

[102] ibid Chapter 3.

[103] ibid Chapter 10 132–134: [...] all law is universal but about some things it is not possible to make a universal statement which shall be correct [...] When the law speaks universally, then, and a case arises on it which is not covered by the universal statement, then it is right, where the legislator fails us and has erred by over-simplicity, to correct the omission-to say what the legislator himself would have said had been present, and would have put into his law if he had known. Hence the equitable is just, and better than one kind of justice—not better

of particular justice, equity has two attributes; it is part of the universal natural law, which calls for application even if not codified (rational element), and it is also a quality of the judge, who must 'correct' a situation otherwise unjust (volitional element).[104]

In Roman times, this Aristotelian duality of 'equity' led to the clear distinction of its two attributes into two separate concepts.[105] The 'rational element' of equity corresponding to a broader interpretation of equity, places the concept within a larger framework of natural law and considers it a source of law and an inseparable element of the adjudication process.[106] The 'volitional element' corresponding to a more limited interpretation of equity, regards the concept within the judicial decision-making elements as a corrective method use by judges to soothe unnecessary distress caused by the application of general principles and specific rules.[107]

It was precisely this separation between the rational and volitional elements of 'equity' that shaped the further evolution of civil and common law traditions. Civil law strictly following Roman tradition places considerable emphasis on the codification process and 'equity' depends on the *conscience juridique*; whereas common law gradually processed the inherited from Roman law duality of 'equity' to reconcile the two attributes into one concept, going back to the original Aristotelian understanding. Accordingly, in common law 'equity' is not merely a means of correction of codified law, but a source of law of its own; a set of principles gradually evolved applicable in cases where the inflexible legal rules were unable to grant relief in new and/or more complicated situations.[108]

Aquinas in the 16$^{th}$ century made a similar categorization of justice to Aristotle: legal, corrective and distributive.[109] Aquinas's argument is based on the three types of relationships: the relations of the parts to the whole, the relations of the parts amongst themselves, and the relations of the whole to the parts. Correspondingly, the three justices are the legal, corrective and distributive; or otherwise, procedural, retributive and distributive.[110]

than absolute justice, but better than the error that arises from the absoluteness of the statement. And this the nature of the equitable, *a correction of law where it is defective owing to its universality* (*emphasis* added).

[104] Rossi (n 98) 27 f.
[105] ibid.
[106] ibid 21.
[107] ibid.
[108] ibid.
[109] Aquinas (n 87) 181 f; Finnis (n 86) 184 f. Aquinas refers to *legal* and *commutative* justice, instead of Aristotle's *universal* and *corrective* justice, but here the Aristotelian words are used to avoid confusion with the different terms.

[110] Mellers in Mellers and Baron (eds) (n 82) 3, refers also to an alternative scheme of categorization of justice in normative, descriptive and prescriptive. Normative research on justice concentrates on rational theories of individual decision making. Descriptive research studies psychological theories of fairness and other morality measures. Finally, prescriptive research on justice examines methods and policies for reducing conflict, encouraging cooperation and resolving differences among individuals or groups.

In modern times, John Rawls proposed a restricted theory of justice for the basic institutions of society.[111] Exploring issues of distributive justice in his famous book, *Theory of Justice* (1971), Rawls developed the concept of justice as fairness. Distributive justice focuses on the production and distribution of society's resources, the assignment or resignment to the members of the society of what they need or what they deserve. Rawls tries to answer how this process is functioning in a way best corresponding justice and accordingly, what is justice, and which are the principles to be followed in order to achieve it, following a welfare liberal approach protecting against classic utilitarianism as proclaimed by John Stuart Mill and Jeremy Bentham. Rawls's ultimate political idea is that all social primary goods (liberty and opportunity, income and wealth, and the bases of self-respect) are to be distributed equally unless an unequal distribution of any or all of these goods is to the advantage of the least favoured.

Rawls adapts the Lockean theory of a social contract to a model society of 'original position', a hypothetical state of nature, where all individuals are free, equal, and rational, with equal rights and duties.[112] These individuals are not aware of their own qualities ('veil of ignorance'), eg of their intelligence, or their social status, and they are to engage in a reasoned debate over principles of justice in order to arrive at a final and binding consensus.

Rawls believes that there are two principles of justice agreed upon at this original position, liberty and wealth. According to the liberty principle, each person has an equal right to the most extensive total system of equal basic liberties compatible with a similar system of liberty for all; according to the wealth principle, social and economic inequalities are arranged to the greatest benefit of the least advantaged and offices and positions in institutions should be open to all under conditions of fair equality of opportunity.[113] Fairness in process corresponds to liberty and implies equal participation of all in the consensus-making; fairness in result reflects wealth, the second principle of justice. For Rawls, fairness is the method/process to find justice, to distribute equally social primary goods among individuals.

### d) International justice: philosophical dimensions and judicial practice

The theories examined in the previous section analyzed justice, fairness and equity applied in the domestic case. These considerations assist in the implementation of the CBD through domestic action, because contracting States

[111] Rawls (n 57) 3 f, 6 f, 73 f, and 94 (see the detailed scheme). Also J Rawls, *Justice as Fairness: A Restatement* (The Belknap Press of Harvard University Press, Cambridge, Massachusetts/London, England, 2001).

[112] Rawls (n 57) 10 f.

[113] ibid 52 f.

are called on to design access and benefit-sharing legislation to flesh out the general CBD provisions.

However, considering the international treaty character of the CBD, there is a further need to explore the international dimensions of 'fair and equitable'. Nowadays, given the great discrepancies between poor and rich countries, there are concerns about international justice and the obligations of the rich towards the poor reflected in several international documents and actions related, for example, to fundamental human rights or humanitarian and economic assistance and capacity-building.

Starting from ancient philosophical discourse, the Aristotelian meaning of 'equity' and its evolution in common law shaped the use of the concept in international legal practice. Grotius, identified as the father of international law, adopted the Aristotelian initial formulation of equity.[114] Compared to the German *Billigkeit* and the French *équité*, the English equity is distinguished for its dynamic in common law jurisdictions. It promotes change and progress and therefore it attracted a lot of attention in the international arena in times of momentous social changes, such as the period following decolonization in the middle of the 20th century.

Aristotelian doctrine serves the interpretation of equity (or fairness or particular justice) in international practice in two ways: first in its distributive form, in the sense of distribution of common resources within global society; and second in its corrective form, as evolved in common law practice and influenced international law practice.

A study dedicated to the 'fair and equitable' objective of the CBD argues that 'equity and fairness' have not been widely used in international law and it is difficult to trace precedents of interpreting them.[115] However, the concept of equity is frequently referred to in recent important international agreements (UNCLOS, FAO International Treaty and CBD) and there have been serious adjudicative attempts for its interpretation. 'Equity' in international law was primarily associated with the distributive claims of developing countries, and the applicability of the principle to the developing law of the sea in the field of maritime boundary delimitation.[116]

In the 1974 'Declaration on the Establishment of a New International Economic Order', and the accompanying 'Programme of Action and the Charter of Economic Rights and Duties of States', the international community affirmed the principles of equity, sovereignty, interdependence, common interest and cooperation among all the states.[117] The common law principle of equity expanded and became a symbol for global justice, and a

114 Brown Weiss (n 48) 35.
115 Byström, Einarsson, and Nycander (n 2).
116 Rossi (n 98) 6.
117 GA Res 3201 (S-VI) and GA Res 3202 (S-VI) of 14 December 1962.

dynamic source of international law.[118] The question of 'what is equitable' has been at the center of debates at the UN General Assembly, the Security Council, multilateral conferences, and academic institutions; but the notion of equitable is also adaptable to the overly political or biased objectives of States and scholars.[119]

Notably, in the law of the sea 'equity' is a main legal concept, invoked and applied to delimit spatial areas that harbour these resources.[120] These resources were not of interest, until technological and geological progress revealed their wealth and a struggle began to govern them. Industrialized countries were able to exploit and expropriate the riches of these resources, and soon the fear was expressed that this would become a customary practice at the expense of developing countries. These facts led to a major transformation in this area of international law.[121] Was this exploitation and expropriation fair? A regime should be negotiated to balance the stakes of all countries on earth, industrialized and developing, to benefit from the common resources. Therefore, equity in modern international law is primarily employed to secure division and use of the *res communis*.[122]

Moving closer to international environmental law, the Brundtland Report puts equity in context with sustainability in the use of natural resources; unsustainable use of resources is also due to unequal distribution of access to them.[123] For food and agriculture, this assertion implies that food security is not only a question of raising average productivity and incomes, but a question of equity in food production and distribution.

With its maritime cases,[124] the International Court of Justice confirmed

[118] Rossi (n 98) 3; TM Franck *Fairness in International Law and Institutions* (Clarendon Press, Oxford, 1995) 48 f.

[119] Rossi (n 98) 4.

[120] United Nations Convention on the Law of the Sea (UNCLOS), adopted on 10 Dec 1982, 21 Int'l Legal Materials 1261 (1982). In UNCLOS Preamble recitals 3 and 4, the States proclaim the sovereignty principle and set the goals of 'the peaceful uses of the seas and the oceans, the equitable and efficient utilization of their resources, the conservation of their living resources and the study, protection and preservation of the marine environment', for the sake of 'realization of a just and equitable international economic order' in the long run. The term is used extensively in UNCLOS Part V (Exclusive Economic Zone), Part VI (Continental Shelf), Part XI (The Area, ie seabed and ocean floor, and its subsoil), and Part XIV (Development and Transfer of Marine Technology).

[121] The great shift in the law of the sea came after the World War II. For the last 400 years, there was only one major doctrinal debate (between Grotius' *Mare Liberum* and Selden's *Mare Calusum* in the 17th century) and one substantive change (van Bynkershoek's three-mile 'canon shot' rule). See Rossi (n 98) 7 and 193 f.

[122] ie a thing that belongs to everyone, cannot be appropriated and must be administered for the benefit of all, in contrast to the *res nullius*, a thing that belongs to no one, and is subject to the appropriation of the state that acts first to exercise control. Rossi (n 98) 196.

[123] Brundtland Report (n 39) 46 f and 141.

[124] North Sea Continental Shelf Cases [1969] ICJ Rep 3 (Judgment of 20 February 1969); Case concerning the Continental Shelf (Libyan Arab Jamahiriya/Malta) [1985] ICJ Rep 13

judicial activism and witnessed the common law influence.[125] The Court based its reasoning on a 'multiple factor analysis', where these factors were a collage of equitable principles, rules, criteria, and methods, as well as 'relevant circumstances', moulded by the Court to determine boundaries. More specifically, in the landmark *North Sea Continental Sea* cases, the judges found the rule-oriented approach impractical and on the basis of the facts they rejected the obligatory equidistance principle,[126] an approach that caused a lot of critique at the time. However, a careful consideration of the Court's decision and argumentation reveals that its approach to delimitation questions conforms to a structure where equity plays a valuable role.[127]

Moreover, in international investment law, the obligation of the parties to provide to each other's investments 'fair and equitable treatment' is a standard for the protection due to foreign direct investment by host countries. There is diversity in the formulation and interpretation of this standard, depending on the context, the negotiating history and the wording of the agreement. Arbitral awards tend to identify 'due diligence' and 'due process' as elements of the normative content of the standard, but their jurisprudence is relatively recent and not uniform, and for this reason it is

(Judgment of 3 June 1985); Case concerning Delimitation of the Maritime Boundary in the Gulf of Maine Area (*Canada v United States of America*) [1984] ICJ 246 (Judgment of 12 October 1984); Case concerning the Continental Shelf (Tunisia/Libyan Arab Jamahiriya) [1982] ICJ Rep 18 (Judgment of 24 February 1982).

[125] Applying the principle of equity according to Article 38(I)(c) of the statute, which allows the Court to refer to 'general principles of law as recognized by civilized nations'.

[126] North Sea Continental Shelf Cases, 1969 ICJ Rep 3 (Judgment of 20 February 1969). These cases arose out of failed attempts by the Netherlands, Denmark, and the Federal Republic of Germany to delimit lateral portions of their respective continental shelves in the North Sea. On the basis of the 1967 Special Agreements for the Submission of a Difference Concerning the Delimitation of the Continental Shelf in the North Sea, the three countries instituted individual proceedings before the ICJ, and the cases were joined for the same nature. The Court was not asked to delimit the boundaries; instead it was asked to distinguish in each case 'what are the applicable principles and rules of international law' (at 2). Its role was thus not adjudicatory, but declaratory. The Netherlands and Denmark held the common position that the equidistance principle proclaimed in Article 6.2 of the 1958 Geneva Convention on the Continental Shelf should be applied; in their view, the fact that Germany had not signed the 1958 Convention did not affect the implementation of the principle, because the equidistance principle formed part of the customary law (27). Germany, however, rejected this idea because of the special circumstances, which deal with the concavity of the North Sea coast line: the application of the equidistance principle would lead to an inequitable result and thus Germany promoted the 'just and equitable share' of the shelf (at 17–20). It further argued that the principle of 'just and equitable share' is a 'basic legal principle emanating from the concept of distributive justice and was legally recognized in all the legal systems, including this of the international community' (see North Sea Continental Shelf Cases, 1968 ICJ Pleadings Vol I, 30; at 31 f. Germany gives examples of State practice). Eventually, the Court rejected the equidistance criterion stating that 'delimitation is to be effected in accordance with equitable principles, and taking into account all relevant circumstances' (at 46).

[127] Rossi (n 98) 216–219).

still difficult to reach a definitive conclusion on the exact meaning of the 'fair and equitable' formula in this context.[128]

The emergence and increasing use of 'equity' in international legal practice is commendable. 'Equity' appears in the sharing of benefits arising from the use of global goods, where it comes to regulate the balanced distribution of income and wealth between the poor and rich of this world. It is used in the sense of resources' distribution and access to opportunities, but also in the sense of compensation: there is a need to compensate in order to counterbalance or to make up for, an injury or a contribution. It thus culminates in a necessary qualification of international economic relations and cooperation; it expresses specific attributes in the developed–developing world partnership and exchange relationships and is associated with the notions of empowerment, participation and capacity-building. Legal recourse to equity introduces elements of fairness into the system, as a procedural aspect in the just distribution of global wealth among States.[129]

It would be interesting to apply Rawls' theory of justice as fairness at the international level, where all primary social goods would be distributed equally among the individuals of the global society, unless unequal distribution would benefit the least favoured in this world.[130] Moreover, a theory that puts the interests of individual human beings at the centre of any settlement of moral questions, would lead to a very progressive and evolutionary international justice system founded primarily on human rights.[131]

However, 28 years after his *Theory of Justice*, Rawls denies this application in the international context and proposes a different moral theory in his *Law of the Peoples*.[132] The fictional forum of 'original position' exists, but the parties are no longer 'free and equal individuals', but 'free and equal peoples',[133] where 'free and equal' are explained as 'well-ordered', having a liberal or decent domestic institutional order. The 'veil of ignorance' is thin enough to allow the distinction of the liberal and decent peoples who are going to join the same agreement, not on the design and functioning of a global institutional order, but on the rules of good conduct, obedience and cooperation.[134]

128 OECD *Fair and Equitable Treatment Standard in International Investment Law* (OECD, Paris, 2004).

129 Franck (n 118) 47–80.

130 In the Theory of Justice an extension of justice as fairness is done by Rawls limited in judging the aims and limits of just war, Rawls (n 57) 331.

131 A Sen, *Development as Freedom* (Oxford University Press, 1999; see also Petersmann (n 84).

132 J Rawls, *The Law of Peoples* (Harvard University Press, Cambridge, 1999).

133 ibid 23; Rawls explains why he chooses 'peoples' and not 'states', but his clarification is not very comprehensible.

134 ibid 3 f and 63.

Pogge rightly criticized Rawls for his different conception of justice at the domestic and international levels and pinpointed the asymmetries between the domestic three-tiered/institutional theory and the international two-tiered/interactional theory.[135] Although Rawls claims a 'duty of assistance' in his international justice theory with the *Law of the Peoples*,[136] he fails to justify it satisfactorily. In this book, Rawls suggests five types of domestic societies: 'reasonable liberal peoples', 'decent peoples', 'outlaw states', 'societies burdened by unfavorable conditions' and 'benevolent absolutisms', and he claims that only the two first types qualify for respect and tolerance and thus should interact with other peoples in order to agree upon certain rules of cooperation. But, following this thinking, affluent States that are not interested in assisting poorer and badly-ordered peoples, would belong to the category of 'outlaw States';[137] and thus they should be denied equal respect and tolerance by the rest. A further shortcoming is that, although in his domestic theory of justice Rawls places the interests of individual human beings above everything else (normative individualism), he rejects this idea in his suggested international justice.

The notion of fairness more specifically in the international legal and institutional system has been analysed by Franck,[138] whose study is constructive in the allocation of global resources. Franck argues that the concept of 'fairness' is composed by legitimacy (procedural aspect) and distributive justice (moral aspect).[139] Fairness is perceived as a specific quality of process, as a necessary condition/attribute of rules governing the allocation of scarce resources within a given community. Fairness functions either as an exception to a non-equitable rule of law ('corrective equity'); as the dominant rule of law by itself ('broadly conceived equity'); or determines the conditions for exploitation and ensures conservation of mankind's common patrimony ('common heritage equity').[140]

Consequently, according to Franck's discourse, when resources belong to States, fairness comes to resolve claim conflicts of the States; whereas, when resources constitute common heritage, a number of principles deriving from this assumption, such as non-ownership of the heritage, shared management, shared benefits, use exclusively for peaceful purposes, and conservation for future generations, steer the management of such resources.[141]

For Franck, fairness is a 'journey or process' leading to a democratic entitlement linked to the questions of self-determination of peoples;[142] it is

135 TW Pogge, 'The Incoherence between Rawls's Theories of Justice' (2004) 73 Fordham Law Review 1744.

136 Rawls (n 132) 5 and 43.

137 ibid 4f.

138 Franck (n 118) 3 f.

139 ibid 7–9.

140 ibid 9 f and 56 f.

141 ibid 75 f.

142 ibid 83 f and 140 f.

related to issues of empowerment of persons and peoples and their participation in the relevant process.

e) The 'fair and equitable' formula in the CBD

Further to the preceding philosophical and judicial examination, 'fairness and equity' in international law and practice are understood as specific qualities for the process and outcome of wealth distribution, and a sort of compensation might be needed to redress a balance among individuals in today's global society. The goal of 'fair and equitable sharing of the benefits arising out of the utilization of genetic resources' is laid down in CBD Article 1 and specified in Article 15. The use of 'fair' and 'equitable' in the CBD is not a tautology; the terms have an emphasizing and complementary function, referring both to the process and the outcome of the distribution of benefits deriving from the use of biological resources.

At the international level, there is a sense of equity as compensation in the interpretation of the CBD. Developing countries, which are the main suppliers of biological material, need to be compensated for the conservation of natural resources and the development of medicinal knowledge, employed by the drug industry. Indigenous communities in particular need to be compensated for their efforts to conserve and sustainably manage genetic resources,[143] and likewise for sharing with the rest of humanity valuable medicinal knowledge they developed over time.

The same can be said for the FAO International Treaty.[144] Farmers who conserve the diversity of crop varieties should be compensated for their contribution of the supply and conservation of plant genetic resources for food and agriculture. The multilateral system aims to correct the imbalance between in situ farmers and rural conservators on the one hand, and breeders and genetic engineers on the other.

At the domestic level, according to the CBD, States enjoy sovereignty over their genetic resources (15.1) and are assigned responsibility for the implementation of the CBD provisions. The CBD leaves the parties room to decide whether the interpretation and implementation of the CBD provisions on access and benefit-sharing will be based on a public or private or mixed approach. Depending on the approach chosen, countries accentuate different expressions of equity, either in procedural terms (procedural

[143] See also BJ Richardson and D Craig, 'Indigenous Peoples, Law and the Environment' in BJ Richardson and D Wood (eds) *Environmental Law for Sustainability* (Hart Publishing, Oxford and Portland, Oregon, 2006) 194; CC Gibson, MA McKean and E Ostrom (eds) *People and Forests: Communities, Institutions, and Governance* (MIT Press, Cambridge, Mass, London/England, 2000).

[144] See FAO International Treaty Articles 9.2, 10 and 13, in view of the farmers' rights and the multilateral system for plant genetic resources and the reports of the Panel of Eminent Experts on Ethics in Food and Agriculture.

equity); in terms of the relationship of the parties (retributive); or in terms of the relationship of the parties to the state (distributive).[145]

Procedural equity deals with the relation of the parts and the whole; it deals with fairness of the procedure to be followed in decision-making. The main point is that all the stakeholders, ie all individuals/entities with rights and interests should participate directly in the decision-making and should have the opportunity to voice their opinions.

Retributive equity relates to the inter partes transactions; it requires a certain degree of reciprocity and balance in the obligations between two or more parties. Contract law goes beyond mere exchange to extra-legal phenomena,[146] and the exchanged values must weigh equally. What is 'equal' is difficult to assess and law cannot thus provide what is fair.[147]

Finally, distributive equity implies that a State should take care that people are properly rewarded for their work, according to the value of their contribution to the social product or according to their efforts or the costs they made for their work. In the application of distributive justice, the State sets some minimum standards and interferes with the contractual freedom. It also tries to face the problem of unjust enrichment, a concept closely associated with the principle of equity, because it is based on the idea that every contracting party gets a fair share.[148]

It is argued here that the 'fair and equitable' CBD objective can be better served, if the public law component is well elaborated, but without interfering too much to private parties' contractual freedom.[149] Although private practice develops interesting models for 'fair and equitable' benefit-sharing,[150] the role of the State remains essential to guarantee human rights respect and coordination of the bioprospecting undertakings.

Finally, reflecting on the maturity of international law today, Rawls' international law theory of justice arrives at non-egalitarian conclusions, because he develops a contract model theory, where the actors are 'peoples'. We are witnessing a shift of international law from a legal order governing diplomatic relations between States to a more complex legal order where non-governmental actors are major players,[151] and where a constant focus

[145] S Peña-Nera, C Dieperink, and H Addink, 'Equitable Sharing Benefits from the Utilization of Natural Genetic Resources: The Brazilian Interpretation of the Convention on Biological Diversity' (2002) 6 Electronic Journal of Comparative Law at http://www.ejcl.org/index.html

[146] PS Atiyah, *Essays on Contract* (Clarendon Press, Oxford, 1986) 1 f.

[147] ibid 329 f.

[148] EJH Schrage, 'Unjust Enrichment: a Historical and Comparative Overview' in EJH Schrage (ed) *Unjust Enrichment and the Law of Contract* (Kluwer International Law, The Hague, New York, 2001) 1.

[149] The Bonn Guidelines (COP Decision VI/24) suggest a public law oriented approach and incorporate some basic public law elements; see Chapter 3 Section III.B.

[150] See Chapter 6.

[151] CG Weeramantry, *Universalising International Law* (M Nijhoff, Boston, 1998/2004).

of international community on the protection of fundamental human rights, in favour of individuals and less powerful communities (indigenous peoples' rights, minorities' rights) is located. Francks' international theory offers useful insights into international justice because democratic principles and self-determination of peoples are central to his argument. Domestic and international benefit-sharing regulatory efforts fail to meet the 'fair and equitable' CBD objective, if they fall short of human rights respect, individual (right to property) and collective (right to self-determination, right to development and indigenous rights) and democratic procedures in the negotiation of bioprospecting agreements[152] related to biogenetic resources located within sovereign States and the distribution of the derived benefits.

### *C. The Interrelationship Among the CBD Objectives*

The analysis of the objectives of the CBD was intended to be as comprehensive as possible. It was first explained that accepting 'conservation' as 'preservation' renders sustainability discussion redundant. 'Conservation' and 'sustainable use' should go together in sustainability policies. 'Sustainability' is currently gaining a lot of attention in environmental law and policy. International practice shows that 'sustainable development' is abandoning its initial economic alignment and culminates in a global ethic calling for a new way of life that is both socially and environmentally 'good'. Finally, a short study of justice/fairness/equity was conducted, in order to illuminate the 'fair and equitable sharing' objective of the CBD, and its meaning at both national and international levels. While discourses of social justice were in the past monopolized within national communities, globalization and integrative States' organizations renders the quest for international justice relevant in the international legal order. It could well be argued that the earlier analysis does not take into consideration the approach to the concepts of sustainability, fairness and equity, in societies other than the Western thought. Indeed, indigenous peoples may have a completely different understanding of what is fair and equal, and this would be very topical for the purposes of the present study. However, these notions were studied with a view to interpreting a text of an international convention, the CBD, drawing on international law, which admittedly draws on Western legal traditions.

The question that arises at this point is: what is the link between the three objectives? The CBD objectives are not separate and independent, but the realization of one of the objectives depends on one or more of the

[152] Prior Informed Consent Procedure and Mutually Agreed Terms according to CBD Article 15.4 and 15.5; see Chapter 3 Section 2.

others. The three objectives are thought to be mutually supported and innately related: there is an intimate link between 'conservation' and 'sustainable use', whereas the present moral international dimensions of 'sustainability' demand 'fair and equitable' distribution of the benefits arising out of the utilization of resources.[153] At the time of signature of the CBD, the 1992 Rio Declaration announced its parties' 'goal of establishing a new and equitable global partnership through the creation of new levels of cooperation among States, key sectors of societies and people',[154] and they proclaim that 'the right to development must be fulfilled so as to equitably meet developmental and environmental needs of present and future generations'.[155]

The theory of intergenerational equity[156] gives further support for the association of sustainability to equity concerns. Intergenerational equity is one of the principal components of sustainability. Present generations are intermediaries between past and future generations in the use of natural and cultural resources, having both the right to enjoy the goods left from the previous generation and the obligation to protect and deliver them to the next one in the situation they received them. Every generation has to conserve the quality of the received resources, the options for their use and the access to them with the objective to achieve the welfare and well-being of any generation.[157] Apart from the this temporal dimension of 'equity' as 'intergenerational', sustainability calls also for 'intragenerational' equity, that is 'equity' among stakeholders, trustees and stakeholders of the same generation.

Unfair and non-equitable sharing of the benefits emanating from the production of medicines produced from genetic resources can lead to inequities among stakeholders of the same generation in the short term, and can deprive future generations of access to millions of species destroyed, and of benefiting from their medicinal qualities in the long term, with implications to conservation and sustainability. 'Sustainable development' is the starting point in pursuing the ways and means to support and maintain the diversity of genetic resources in parallel with economic development. To face global welfare problems, resources that support developing countries and especially their local and indigenous communities need to be generated. Research and development in medicines from genetic resources and use of associated traditional knowledge can promote such development concerns,

[153] The link between equity and sustainability was also demonstrated with the Brundtland Report (n 39) 46 f and 141 f.

[154] Rio Declaration Preamble.

[155] Rio Declaration para 3. See also the International Law Association Principles of International Law Related to Sustainable Development (ILA New Delhi Declaration of 2002).

[156] Brown Weiss (n 48) 1 f and 193 f.

[157] See also Rawls (*Theory of Justice* n 57) 251 f.

but at the same time they must be accompanied by equity and social justice considerations, respect for social self-determination, and cultural diversity, and maintenance of ecological integrity and conservation.

Unfair negotiation processes, either because not all stakeholders are sufficiently represented, or because the procedures are not sufficiently democratic; and inequitable distribution of the benefits arising from the use of the biogenetic resources, lead to unsustainable management and devastation of the resources in pursuit of more lucrative economic activities. Taking into account that most developing countries host vast biological wealth and that indigenous peoples have for centuries developed sustainable management systems, fair and equitable benefit-sharing in bioprospecting supports conservation and sustainable use and promotes development efforts.

## IV. SUMMARY AND CONCLUSION

The CBD model reminds us that biodiversity management is more than a conservation exercise. In particular, the third objective stresses the importance of fairness in process and equity in biodiversity outcomes, recognizing that, at present, the costs and benefits of biodiversity are not shared equally throughout society.

Bioprospecting is not a new activity, but there is a need to place it into a broader political and economic context, especially a neo-liberal, capitalistic one, that integrates four interlocking factors: market and economic rationales; rapid and broad biotechnological changes; the growing interest of the drug industry to link bioprospecting profits to environmental protection; and the international efforts for the harmonization and standardization of intellectual property and biodiversity regimes.[158] Reshaping the bioprospecting legacy is not an easy task: operations of companies that try to promote access and benefit-sharing practices face the opposition from the pharmaceutical industry; the lack of effective domestic institutional and legal infrastructure favours the free acquisition of potentially valuable resources by any means; and intellectual property agencies are not ready to deal with indigenous communities and traditional knowledge.

Source countries should be assisted to better understand and value their fundamental dependence on ecosystem services and for this reason, non-monetary benefits are critical. In a row, training and education will raise the opportunity costs of land conversion by increasing the productivity of their

[158] MK Dorsey, 'The Political Economy of Bioprospecting in Amazonian Ecuador—History, Political Economy, and Knowledge' in SR Brechin et al (eds) *Contested Nature: Promoting International Biodiversity Conservation With Social Justice in the Twenty-first Century* (State University of New York, Albany, 2003) 137.

labour applied elsewhere, eg in adding value to the genetic resources. Basic investments in education and rural institutional and human resources' infrastructure induce private investment in value-added manufacturing, processing, and distribution.

Given that biogenetic resources of medicinal importance and traditional medicinal knowledge are often found in indigenous lands, indigenous communities qualify as stakeholders in the bioprospecting undertakings: they should participate in the negotiations of the agreements, grant prior informed consent and receive compensation. Otherwise, bioprospecting fails to meet the CBD goals. Unfortunately, in many States the indigenous peoples' position is very disadvantaged and their rights are not fully respected and protected. The expansion of human rights in international law with regard to indigenous peoples calls for domestic policies that flesh out these international norms.[159]

Besides the domestic protection of indigenous communities, well-defined property/land tenure rights promote the CBD objectives, because they facilitate bioprospecting transactions over resources and boost the functioning of the bioprospecting market. Property rights over lands and natural resources should not be realized exclusively in the way they have been perceived so far, but flexibility should allow for recognition of clear customary property rules developed by local and indigenous societies. Moreover, great challenges arise with regards to property rights over genetic material and medicinal knowledge.[160]

[159] See Chapter 7 Section III.
[160] See Chapter 7 Section II.

# CHAPTER 3

# *Access and Benefit-Sharing in the CBD–Bonn Guidelines*

## I. INTRODUCTION

This chapter explains the more specific conditions set by the CBD on access and benefit-sharing: State sovereignty, mutually agreed terms, prior informed consent, and benefit-sharing.

The parties to the CBD, aware of the broad and vague character of the CBD provisions, started rather early the efforts to facilitate the fulfilment of these provisions; through information gathering on domestic implementation and on developed private practices, through collaboration and coordination with other intergovernmental organizations, and assignment of access and benefit-sharing, and traditional knowledge, specific work to expert groups.

The Bonn Guidelines, adopted with Decision VI/24 from the parties at their sixth meeting, constitute an essential and very detailed guidance for domestic implementation of access and benefit-sharing in both the provider and user countries. The Guidelines contain critical instructions for the furtherance of the CBD, but are of a non-binding and aspirational character. The positions of parties during the negotiation of the text reflect the tensions and the discord at the international level and signal the difficulties for the negotiation of a single treaty on access and benefit-sharing.

Finally, the access and benefit-sharing implementation options, namely international, domestic (national and regional), and private practices, are investigated laying the groundwork for a more thorough analysis in the next chapters.

## II. ACCESS TO GENETIC RESOURCES AND BENEFIT-SHARING IN THE CBD

Article 15 of the CBD is the foundation article laying down the rules for regulating access to genetic resources and benefit-sharing. Access rules to genetic resources regulate the procedures for obtaining the permission to do research on or with biological resources. Granting access to genetic resources does not necessarily imply sharing the benefits of subsequent use of genetic resources, even if the Article calls on negotiating parties to do so 'with the aim of sharing in a fair and equitable way the results of research and development and the benefits arising from the commercial and other utilization of genetic resources'.[1] Clearly, it is recommended that access agreements should provide for a minimum protection of the commercial and non-commercial interests of the Contracting Parties.

The CBD rules governing access encompass all types of biological and genetic resources irrespective of whether they are wild or domesticated, of animal, plant, or microbial origin, situated in private or public land, or water. However, there are some important exceptions. First, access rules do not apply to genetic resources collected prior to the CBD's entry into force neither in a particular state, nor to *ex situ* collections. Secondly, facilitated access to some plant genetic resources used in food and agriculture are subject to the specific regulations set out by the FAO International Treaty of the Food and Agriculture Organization.[2] Finally, the Conference of the Parties to the CBD interpreted the access and benefit-sharing provisions in such a way as to exclude human genetic resources.[3]

Above all, Article 15 specifies the rights and obligations of States related to the access to genetic resources and their ultimate utilization. The article authorizes States to provide for treatment of and access to genetic resources,[4] under the condition that the State provisions facilitate 'environmentally sound uses by other Contracting Parties'. Moreover, they must not 'impose restrictions that run counter to the objectives of this Convention'.[5]

Thus, the first two paragraphs of Article 15 establish a balance between these rights and obligations. On the one hand, States have the right to determine the conditions for access to the genetic resources in their territory; on the other hand, they have the obligation to facilitate access of other parties to the genetic resources under their jurisdiction. Furthermore, the CBD states that access 'shall be on mutually agreed terms'[6] and 'subject to prior informed consent'.[7] The States are then obliged to set the specific terms and conditions specifying these CBD objectives.

These principles provide an international legal basis for a *quid pro quo* arrangement whereby CBD Contracting Parties grant access to genetic resources in exchange for a fair and equitable share of the benefits derived from their use. International law and national laws are key tools in achieving this goal. It should be noted that the provisions contained in Article 15 are also linked to the provisions on access to and transfer of technology,[8] exchange of information,[9] technical and scientific cooperation,[10] the

1 CBD Article 15.7.
2 International Treaty on Plant Genetic Resources, FAO Res. 3/2001 (Rome, 3 Nov 2001).
3 COP Dec. II/11 para 2.
4 CBD Article 15.1.
5 CBD Article 15.2.
6 CBD Article 15.4.
7 CBD Article 15.5.
8 CBD Article 16.
9 CBD Article 17.
10 CBD Article 18.

handling of biotechnology and the distribution of its benefits,[11] and financial resources and financial mechanisms.[12]

## A. Access to Genetic Resources

Access to genetic resources is premised on three fundamental principles: State sovereignty over genetic resources, mutually agreed terms (MATs), and prior informed consent (PIC).

### 1. *State sovereignty and the legal status of genetic resources (CBD, Article 15.1)*

Both Article 15 of the CBD and Agenda 21 clearly establish the sovereign right of States over resources found within their borders but also stress the duty to conserve biodiversity and promote sustainable development.[13] Agenda 21 in fact states that States 'have the sovereign right to exploit their own biological resources pursuant to their environmental policies'.

Before the CBD entered into force,[14] genetic resources were considered part of the common heritage of mankind. Importantly, placing genetic resources in the 'common heritage of mankind' has a symbolic meaning: it denotes the importance of these resources for all humanity and should not be incompatible with the exercise of State sovereignty, especially when these resources are found within the territory of a country, such as the biogenetic resources, which differ from classic global commons resources that are not clearly circumscribed by national borders, such as the high seas and airspace.[15] Hence, CHM should not necessarily imply that such resources are accessible to and usable by anyone without restrictions. However, in practice, genetic resources ranked in the CHM are treated as commons, resulting in overuse and extinction.[16]

To correct this fallacy, the international community attached the same importance to the conservation and sustainable use of resources by replacing the CHM in the CBD with the 'common concern' of mankind and reaffirmed the authority of the State. In turn, States have accepted their increased responsibility to regulate and manage access and benefit-sharing. This recent emphasis in exercising control over biological resources contrasts markedly to past approaches largely due to the fact that States

[11] CBD Article 19.1 and 19.2.

[12] CBD Articles 20 and 21.

[13] CBD Articles 6, 8, and 10.

[14] See Chapter 2 Section II.

[15] See G Dutfield, *Intellectual Property, Biogenetic Resources and Traditional Knowledge* (Earthscan, London, 2004) 10; L Glowka, *A Guide to Designing Legal Frameworks to Determine Access to Genetic Resources* (IUCN, Gland, Cambridge, Bonn, 1998) 4.

[16] For the economic consequences of this perception, see Chapter 1 Section III.C.

now feel the pressure of scarcity, while not long ago natural assets were perceived as abundant and conservation measures unnecessary.[17]

At the same time, the FAO International Undertaking on Plant Genetic Resources[18] was revised in 1991 to align itself to the CBD and thus stated that 'the common heritage of mankind [...] is subject to the sovereignty of the States over their plant genetic resources'. The CBD highlights States' sovereign rights over genetic resources within their borders, but it cannot interfere with the property status of these resources within the States' borders.[19] Sovereignty differs from property: sovereign rights are the rights wielded by the State by regulating its citizens' behaviour by passing and enforcing laws and by being recognized by the international community as an independent state. Property rights grant their holder the right to use and enjoy and own the thing or idea over which s/he has rights and are subject to national legislation

Hence, it is up to the national legal framework to resolve the legal status of genetic resources. Defining the ownership interest issue in access and benefit-sharing regulation is important in order to clarify who is entitled to negotiate, grant the PIC, and share in the benefits derived from the use of genetic resources.

Exercising sovereignty over genetic resources is not the same as exercising sovereignty over natural resources, such as mineral ores or hydrocarbons. Genetic resources are renewable resources and there is a notable disadvantage for the source countries to physically control the access to and flux of genetic resources. The control should not be insurmountable, if the countries within the same region that share similar biological wealth coordinate among themselves and adjust the political and normative process.

### 2. *Mutually Agreed Terms—MATs (CBD, Article 15.4)*

According to Article 15.4 of the CBD, the Contracting Parties need to negotiate and agree on the terms of the access agreement (sometimes called the 'material transfer agreement' or 'academic commercial research agreement') that authorizes access to genetic resources, controls subsequent use, and establishes the type of benefits to be returned. If the provider country claims ownership over the genetic resources then it is up to the State to negotiate

[17] AC Kiss, 'La notion de patrimoine common de l'humanité (1983) 194 *Recueil des Cours—Collected Courses of the Hague Academy of International Law, Académie de droit international* 103; W Lesser, *Sustainable Use of Genetic Resources under the Convention on Biological Diversity: Exploring Access and Benefit-Sharing* (CABI Publishing, New York, 1997), 13.

[18] International Undertaking Article 1.

[19] Glowka (n 15) 4 and 31; J Chaves, 'The Andean Pact and Traditional Environmental Knowledge' in NP Stoianoff (ed) *Accessing Biological Resources—Complying with the Convention on Biological Diversity* (Kluwer Law International, New York, 2004) 244.

the MATs and enter into the agreement. Even if the State is not the owner, it can still exercise a certain amount of control over agreements on genetic resources within its jurisdiction stipulated between the user and other private or communal parties by virtue of sovereignty rights.

### *3. Prior Informed Consent—PIC (CBD Article 15.5)*

Article 15.5 of the CBD lays down the conditions regulating access to genetic resources in accordance with the PIC of the provider of the genetic resources. The potential user must inform the provider country of the intention to access and use genetic resources within the jurisdiction of the country before such activities have begun, and obtain consent to do so. The potential user must, therefore, provide information on the kind and quantity of the resource desired, as well as the timeframe and purpose of the activities. In order to facilitate this process, States must legally define the steps involved in the PIC procedure and designate a supervising administrative body to handle it. The procedure must be clear, ensure that sufficient information is communicated, and pave the way for mutually agreed terms. Finally, PIC procedures should not just relate to access requirements, but comprise informed consent for benefit-sharing, thus fulfilling this CBD objective.[20]

The CBD considers the development of a legal PIC procedure as a prerequisite 'unless otherwise determined by the Party'. This exception refers to any access procedures (usually related to research, collection, import and export permits) that some countries may already have had in place before the CBD came into force. Although these pre-existing provisions may already be clear, detailed, and adequately meet the CBD requirements, it is recommended that they be reviewed and revised to take into account the benefit-sharing aspect.

## *B. Benefit-Sharing and Intellectual Property Rights*

The main concern of the CBD is not so much to set the ground for access procedures, but to ensure that benefits arising from the utilization of natural resources are shared and to address global equity considerations. This is achieved by linking the issues of access with that of the sharing of direct and indirect, monetary and non-monetary benefits. Apart from Articles 1, 15.6 and 15.7, benefit-sharing is also dealt with in Articles 16, 19.1 and 19.2. Through these provisions, the CBD calls for Contracting

[20] LA Firestone, 'If You Say Yes, I Say No: Defining Prior Informed Consent under the Convention on Biological Diversity' (2003) 16 Georgetown International Environmental Law Review, 171 and Glowka (n 15) 9 and 55.

Parties to: carry out scientific research based on shared genetic resources,[21] share the results of research and development efforts and the benefits deriving from the commercial or other use of genetic resources,[22] engage in technology transfer,[23] ensure the participation of provider countries in biotechnological research,[24] and promote and advance priority access, on a fair and equitable basis, to the results and benefits arising from biotechnological research activities.[25]

Prior to the CBD and the introduction of specific international access and benefit-sharing conditions, the allocation of benefits deriving from the trade of products developed from genetic material was governed solely by international intellectual property rights.[26] This meant that the biotechnology industry patented the products it developed on the basis of biogenetic material and could reap all the benefits by the commercialization without having to recompense the country of origin.[27] To an extent, the CBD remedies this imbalance by upholding the right of all stakeholders to claim monetary benefits that arise from a (possibly patented) product derived from genetic resources. Stakeholders include the industry, the provider country and, where applicable, indigenous peoples and other associated individuals/communities.

The CBD does not dictate benefit-sharing rules, but offers guidelines to States on how to achieve fair and equitable sharing. Moreover, the CBD does not specify what the benefits to be shared are; these are to be assessed on a case-by-case basis and depend very much on local particularities. The way in which the CBD's principles are adopted and adapted by national legislations is, therefore, crucial in promoting effective and context-specific benefit-sharing provisions.

Another limitation of the CBD rests on the fact that the Convention is an intra-governmental document that attempts to regulate access and benefit-sharing agreements stipulated between private sector representatives and provider countries. Thus, it is also up to the industry to define what is 'fair and equitable'. Unless these matters are well regulated in national law, conflicts may arise between the user State's commitment to international obligations and its duty to protect the interests of its private sector.[28]

The Bonn Guidelines provide for a more complete framework of benefit-sharing by outlining procedures, listing the types of benefits that can be

[21] CBD Article 15.6. [22] CBD Article 15.7. [23] CBD Article 16.3.

[24] CBD Article 19.1. [25] CBD Article 19.2.

[26] See Chapter 2 Section II.B.

[27] B Dhar and RV Anuradha, 'Access, Benefit-Sharing and Intellectual Property Rights' (2004) 7 The Journal of World Intellectual Property 597; T Cottier, 'The Protection of Genetic Resources and Traditional Knowledge: Towards More Specific Rights and Obligations in World Trade Law' (1998) 1 Journal of International Economic Law 555.

[28] CBD Article 15.7.

shared, and the distribution, timing, and mechanisms that allow States to implement the Convention's principles through national legislative measures.

### C. *Indigenous Peoples*

The CBD is mindful of the fact that in many countries rich in natural resources, indigenous and local communities have, for many generations, been the stewards of biodiversity and developed biodiversity management techniques. As such, they cannot be left at the margins of international access and benefit-sharing discussions. In recognition of the contributions to conservation and to the sustainable use of biological diversity made by the traditional practices of many local peoples, Article 8j of the CBD contains a provision that encourages, but does not require, the equitable sharing of benefits arising from the utilization of the knowledge, innovations, and practices of indigenous and local communities. In this context, it is clear that rules regulating indigenous peoples' relationship with the State and with the international community come into play. [29]

A corollary to Article 8j is Article 10c, which calls for the protection and encouragement of 'customary use of biological resources in accordance with traditional cultural practices that are compatible with conservation or sustainable use requirements'.

## III. THE CONFERENCE OF THE PARTIES AND THE BONN GUIDELINES[30]

Several countries have laws related to access to biological resources,[31] or they are planning to regulate access to ensure benefit-sharing. These laws typically focus on research, collection, or exportation. A key issue is whether they should be made to ensure benefit-sharing in the use of genetic resources embodied in the biological resources. The CBD formalized politically and legally the possibilities for the States' right to capture the benefits

[29] See United Nations Declaration on the Rights of Indigenous Peoples, Doc A/RES/61/295 adopted by the General Assembly on 13 September 2007 and the International Labour Organization Convention No 169 on Indigenous and Tribal Peoples, adopted on 27 June 1989, 28 Int'l Legal Materials 1384 (1989); also, Economic and Social Council, Commission on Human Rights—Sub-commission on the Promotion and Protection of Human Rights, 56th Session, *Final Report of the Special Rapporteur, Erica-Irene A Daes*, Prevention of Discrimination: Prevention of Discrimination and Protection of Indigenous Peoples, Indigenous Peoples' Permanent Sovereignty over Natural Resources, Doc E/CN.4/Sub.2/2004/30, 13 July 2004.

[30] As of December 2008. A previous version appeared at A Smagadi, 'The Impact of the Convention on Biological Diversity on the Utilization of Plant Genetic Resources and Benefit-Sharing' (2006) 6 Yearbook of European Environmental Law 120.

[31] See Chapter 5.

derived from genetic resources. The work of the CBD Secretariat on access and benefit-sharing issues was not only initiated in order essentially to put in place the CBD's objectives and tackle the so-called 'biopiracy' cases in the commercialization of biogenetic resources and associated traditional knowledge, but, also because of the realization that genetic resources in the form of agriculture, horticulture, pharmaceuticals and cosmetics, are potential markets with immense profits.

Today, several years now into the CBD's implementation, we observe that the existing or emerging national legal frameworks determining access to genetic resources have not fully 'materialized'. Developing States are usually rushing to draft and enact legislation before the issue has been carefully examined so that the regulatory frameworks suffer from over-simplicity and lack of clarity. This trend is rather disturbing as the fundamental guiding principles of the CBD are not properly performed. Suggested solutions here relate first of all to capacity building. Source countries need assistance to plan better and develop opportunities to add value to the material they provide.

Another crucial aspect to ensure benefit-sharing is to shift part of the burden from providers of genetic resources to users and their governments. The legal basis for shifting the burden is CBD Article 15 (7), which provides that '*each* Contracting Party shall take legislative, administrative or policy measures...with the aim of benefit-sharing' (emphasis added). This indicates that not only providers of genetic resources, but users of genetic resources are also obliged to take action. To this end a continuum of measures by government and the private and professional sector can be envisioned. Governments' activities could involve adjustment of intellectual property regimes to require the identification of the country of origin. Alternatively, the application process of a patent could be conditioned upon confirmation of PIC.

Finally, measures creating incentives for the industry could be created to secure PIC and benefit-sharing. Private and professional sector initiatives might include the establishment of codes of conduct or best practices definition.

### *A. The Conferences of the Parties on the Access to Genetic Resources and Benefit-Sharing*

The CBD Conference of the Parties (COP) identified early the interpretation and implementation difficulties of access and benefit-sharing, and increasingly lavished attention on this issue. This section narrates the historical treatment of access and benefit-sharing by the Parties to the CBD.

The work started with the COP 2 in Jakarta (6–17 November 1995), where the Parties considered the compilation of existing legal experience on access to genetic resources and benefit-sharing, on intellectual property rights, and access to and transfer of technology that makes use of genetic resources; the information was solicited by governments and relevant international organizations.[32] On this basis, at the COP 3 in Buenos Aires (4–15 November 1996), information exchange was underscored and Parties were asked to express their views on possible options for development of implementation of Article 15.[33]

At the COP 4 in Bratislava (4–15 May 1998), equitable sharing of benefits arising out of the utilization of genetic resources was for the first time addressed by the Parties as a separate agenda item. The issue was looked at from three angles: (a) measures to promote and advance the distribution of benefits from biotechnology in accordance with Article 19; (b) means to address the fair and equitable sharing of benefits arising out of genetic resources; and (c) compilation of views of the Parties on possible options for developing national legislative, administrative or policy measures, as appropriate, to implement Article 15.[34] By Decision IV/8[35] on access and benefit-sharing, the COP decided to establish a regionally balanced Panel of Experts appointed by governments, composed of representatives from the private and public sectors, as well as representatives of indigenous and local communities, charged with the task to define a common understanding of basic concepts and directed to explore all options for access to genetic resources and benefit-sharing (guiding principles, guidelines and codes of best practice).

At its fifth meeting in Nairobi (15–26 May 2002), the COP started to look more closely at the matter. It was decided to reconvene the Panel of

[32] CBD Report by the Secretariat, *Access to Genetic Resources and Benefit-Sharing: Legislation, Administrative and Policy Information*, Conference of the Parties to the Convention on Biological Diversity, Second Meeting, Jakarta, 6–17 November 1995, Doc. UNEP/CBD/COP/2/13, 6 October 1995; and CBD Secretariat Note, *Intellectual Property Rights and Transfer of Technologies Which Make Use of Genetic Resources*, Conference of the Parties to the Convention on Biological Diversity, Second Meeting, Jakarta (6–17 November 1995), Doc UNEP/CBD/ COP/2/17, 6 October 1995.

[33] CBD Report by the Executive Secretary, *Access to Genetic Resources*, Conference of the Parties to the Convention on Biological Diversity, Third Meeting, Buenos Aires, Argentina (4–15 November 1996), Doc UNEP/CBD/COP/3/20, 5 October 1996.

[34] CBD Report by the Executive Secretary, *Measures to Promote and Advance the Distribution of Benefits from Biotechnology in Accordance with Article 19*, Conference of the Parties to the Convention on Biological Diversity, Fourth Meeting, Bratislava (4–15 May 1998), Doc UNEP/CBD/COP/4/21, 16 February 1998; CBD Note by the Executive Secretary, *Addressing the Fair and Equitable Sharing of the Benefits Arising out of Genetic Resources: Options for Assistance to Developing Country Parties to the Convention on Biological Diversity*, Conference of the Parties to the Convention on Biological Diversity, Fourth Meeting, Bratislava (4–15 May 1998), Doc UNEP/CBD/COP/4/22, 9 February 1998; and CBD Note by the Executive Secretary, *Review of National, Regional and Sectoral Measures and Guidelines for the Implementation of Article 15*, Conference of the Parties to the Convention on Biological Diversity, Fourth Meeting, Bratislava (4–15 May 1998), Doc UNEP/CBD/COP/4/23, 19 February 1998, respectively.

[35] Conference of the Parties to the Convention on Biological Diversity, Fourth Meeting, Decision IV/8 on *Access and Benefit-Sharing*, Bratislava (4–15 May 1998), para 3.

Experts on Access and Benefit-Sharing[36] to conduct further work on outstanding issues from its first meeting; to assess user and provider experience, to study complementary options, and identify approaches to the involvement of stakeholders in accessing genetic resources and benefit-sharing processes.[37] In addition, the COP decided to establish the Ad Hoc Open-ended Working Group on Access and Benefit-Sharing, a subsidiary body to the COP. The mandate of the Working Group was to develop guidelines for submission to the COP 6 and assist Parties to the CBD and stakeholders in addressing issues related to access and benefit-sharing, such as PIC, MATs, stakeholders' roles and responsibilities, *in situ* and *ex situ* conservation and sustainable use, and benefit-sharing mechanisms, such as technology transfer and joint R&D. The work of the Working Group would also involve means to ensure the 'respect, preservation and maintenance of knowledge, innovations and practices of indigenous and local communities embodying traditional lifestyles' relevant for the conservation and sustainable use of biological diversity, taking into account work by the WIPO on intellectual property rights issues.[38]

The Working Group met in Bonn, Germany from 22–26 October 2001.[39] The meeting developed a series of guidelines on access and benefit-sharing, which were then considered and adopted by the COP 6, in The Hague (7–19 April 2002), where access and benefit-sharing was one of the priority themes. The adoption of the 'Bonn Guidelines on access to genetic resources and the fair and equitable sharing of the benefits arising from their utilization' was the major achievement of the COP 6,[40] because together with the FAO International Treaty, they shape the growing framework of rules on the access and benefit-sharing legislation.

[36] At its first meeting, held in San Jose, Costa Rica, in October 1999, the panel of experts addressed options for access and benefit-sharing on mutually agreed terms and reached broad conclusions on: PIC, mutually agreed terms, information needs and capacity-building; see CBD Note by the Executive Secretary, *Options for Access and Benefit-Sharing Arrangements*, Panel of Experts on Access and Benefit-Sharing, First Meeting, San Jose (4–8 October 1999), Doc UNEP/CBD/EP-ABS/2, 3 September 1999.

[37] Conference of the Parties to the Convention on Biological Diversity, Fifth Meeting, Decision V/26 on *Access to Genetic Resources*, Nairobi (15–26 May 2000), para 10.

[38] COP Dec V/26 para 11.

[39] Input into the Working Group was provided at the second meeting of the Experts' Panel on ABS (Montreal (19–22 March 2001), which addressed user and provider experiences in ABS and the involvement of stakeholders in ABS processes, see CBD *Report of the Panel of Experts on Access and Benefit-Sharing on the Work of its Second Meeting*, Ad Hoc Open-ended Working Group on Access and Benefit-Sharing, First Meeting, Bonn (22–26 October 2001), Doc UNEP/CBD/WG-ABS/1/2, 9 April 2001.

[40] Conference of the Parties to the Convention on Biological Diversity, Sixth Meeting, Decision VI/24 on *Access and Benefit-Sharing as Related to Genetic Resources*, The Hague (7–19 April 2002); COP Dec VI/24 includes three parts: Part A. Bonn Guidelines on Access to Genetic Resources and Fair and Equitable Sharing of the Benefits Arising out their Utilization, Part B. Other approaches, including the development of an action plan for capacity-building, and Part C, Role of Intellectual Property Rights.

The Parties called for the Working Group to reconvene and advise on the progress of implementation of the Bonn Guidelines, and considered the development of an Action Plan for capacity building.[41] Finally, on the role of intellectual property rights in access and benefit-sharing arrangements[42] the COP encouraged Parties to disclose the country of origin of genetic resources and of related traditional knowledge, and acknowledged that further work is needed in coordination with WIPO and other relevant international organizations.

Following the call for action at the 2002 Johannesburg Summit, the COP at its seventh meeting, in Kuala Lumpur (9–20 February 2004), stressed the negotiation of an international regime on access to genetic resources and benefit-sharing, including the interpretation of CBD Article 8(j), by the Working Group on ABS according to agreed terms of reference for such negotiation,[43] including the process, nature, scope and elements for consideration in the elaboration of this regime. The feasibility, practicality, and costs of measures supporting compliance with prior informed consent of the provider of genetic resources and with mutually agreed terms with the user, were also considered.[44] The parties developed a framework to enhance the evaluation of achievements and progress in the implementation of the Strategic Plan towards the 2010 target.[45]

Following the mandate of the COP 7 and the terms of reference, the Working Group on ABS held its third and fourth meetings in Bangkok from 14–18 February 2005, and in Granada, Spain, from 30 January–3 February 2006, respectively. A number of issues revealed the complexity of the elaboration of an international regime on access and benefit-sharing: the difficulty to develop a common vision regarding the nature of the regime or even its necessity, the linkages of ABS with intellectual property rights, the diverse approaches and the costs for an international certificate of origin/source/legal provenance.[46] A map was further developed to identify

[41] COP Dec VI/24 Part B.

[42] COP Dec VI/24 Part C.

[43] The terms of reference for the negotiation of the international ABS regime were based on recommendations prepared by the Working Group at its second meeting in December 2003, which were submitted to the COP 7 (see the CBD *Report of the Ad Hoc Open-ended Working Group on Access and Benefit Sharing on the Work of its second Meeting*, Conference of the Parties to the Convention on Biological Diversity, Seventh Meeting, Kuala Lumpur (9–20 February 2004), Doc UNEP/CBD/COP/7/6, 10 December 2003).

[44] Conference of the Parties to the Convention on Biological Diversity, Seventh Meeting, Decision VII/19 on *Access and benefit-sharing as related to genetic resources (Article 15)* (9–20 February 2004), COP Dec VII/19.

[45] Conference of the Parties to the Convention on Biological Diversity, Seventh Meeting, Decision VII/30 on *Strategic Plan: Future Evaluation Progress* (9–20 February 2004).

[46] CBD *Report of the Ad Hoc Open-ended Working Group on Access and Benefit-Sharing on the Work of its third Meeting*, Ad Hoc Open-ended Working Group on Access and Benefit-Sharing, Third Meeting, Bangkok (14–18 February 2005), Doc UNEP/CBD/WG-ABS/3/7, 3 March 2005; and CBD *Report of the Ad Hoc Open-ended Working Group on Access and Benefit-Sharing on the Work of its second Meeting*, Conference of the Parties to the

and analyse the gaps in international instruments and indicate ways to address them.

The Working Group made recommendations at the COP 8, Curitiba, Brazil (20–31 March 2006),[47] where it was clear that the negotiations of an international regime in the framework of the COPs and the Working Groups' activities was still far from advanced: the level of indigenous peoples' participation was unsatisfactory, while the debate on the international certificate of origin/source/legal provenance and on the disclosure of origin in intellectual property rights applications remained unresolved. The Parties extended their mandate to the Working Group for the elaboration and negotiation of an international regime on access and benefit-sharing.[48]

Continuing its deliberations, the recommendations of the Working Group fifth and sixth meetings in Montreal, Canada from 8–12 October 2007, and in Geneva, Switzerland, from 21–25 January 2008, respectively,[49] were submitted to the COP 9 for its consideration. At this more recent meeting, the Parties agreed on a roadmap to complete the elaboration and negotiation of the international regime on access and benefit-sharing before the COP 10, to be held in Nagoya, Japan, in 2010,[50] and agreed on the main components of the anticipated regime.[51] The Working Group is expected to meet three times before the COP 10, and will be assisted by three groups of legal and technical experts: i) on compliance; ii) on concepts, terms, working definitions and sectoral approaches; and iii) traditional knowledge associated with genetic resources.

The adoption of a roadmap for the negotiation of an international regime on ABS with the 2010 deadline has been the greatest achievement of the Parties since the adoption of the Bonn Guidelines in 2002. The deliberations of the Working Group and the COP 10 will be particularly critical on the matter. The activities of the Working Group on Article 8(j),[52] should

Convention on Biological Diversity, Eighth Meeting, Curitiba, Brazil (21–31 March 2006), Doc UNEP/CBD/COP/8/6 (15 February 2006).

47 ENB, 'Eighth Meeting of the Conference of the Parties to the Convention on Biological Diversity (20–31 March 2006), Monday 3 April 2006 (2006) 9 Earth's Negotiations Bulletin 363.

48 Conference of the Parties to the Convention on Biological Diversity, Eighth Meeting, Decision VIII/4 on *Access and Benefit-Sharing* (20–31 March 2006).

49 CBD *Report of the Ad Hoc Open-ended Working Group on Access and Benefit-Sharing on the Work of its fifth Meeting*, Ad Hoc Open-ended Working Group on Access and Benefit-Sharing, Fifth Meeting, Montreal (8–12 October 2007), Doc UNEP/CBD/WG-ABS/5/8, 15 October 2007; and CBD *Report of the Ad Hoc Open-ended Working Group on Access and Benefit-Sharing on the Work of its sixth Meeting*, Conference of the Parties to the Convention on Biological Diversity, Ninth Meeting, Bonn (19–31 May 2008), Doc UNEP/CBD/COP/9/6, 13 January 2008.

50 Conference of the Parties to the Convention on Biological Diversity, Ninth Meeting, Decision IX/12 on *Access and Benefit-Sharing* (19–30 May 2008).

51 ibid Annex A.

52 The activities of this Group were omitted from the account of access and benefit-sharing

also be considered. Until the adoption of a new text, the Bonn Guidelines adopted at COP 6, are the main instrument providing guidance on how access and benefit–sharing is to be implemented.

## *B. The Breakthrough of the COP 6: The Bonn Guidelines*

### *1. The general provisions*

Identical to the CBD, the goal of the Bonn Guidelines is the conservation and sustainable use of biodiversity. It is a voluntary[53] agreement, a non-binding legal instrument that focuses on the enhancement of national legislation. It provides instructions for the development of the latter to smooth access to genetic resources and to guarantee that the benefits of any commercialization, research or development are fairly distributed between the users and the owners of the genetic resources.[54] It thus promotes a bilateral system, where third parties' participation is based on the original bilateral agreement.

The preamble states that the Bonn Guidelines are intended to be evolutionary. This wording is rather promising, because it opens the path to a fully-fledged treaty on access and benefit-sharing in the future. The development from the International Undertaking, a voluntary agreement, to the FAO IT, is a recent example that gives hope for similar prospects for the Bonn Guidelines.

The Guidelines are intended to be simple and easy to use,[55] but actually they are quite detailed. The reason is entrenched in its non-binding nature. Governments are more generous and inclined to accept elaborate rules when these rules are only meant as a guide.

Despite their elaborate character, the Guidelines include many areas that have still not been agreed upon: the use of terms, the measures for verification of compliance with the Guidelines' provisions on PIC, and how to deal with genetic resources' derivatives (offshoots that may have started with the genetic resources but have been synthesized into another form and then commercialized, like pharmaceuticals).

The use of derivatives is in many ways associated with the important issue of intellectual property rights and thus with the query whether components of a genetic resource are actually patentable. The Bonn Guidelines do not set any specific restrictions on intellectual property rights for genetic resources. The role of intellectual property rights in the implementation of access and benefit-sharing arrangements do not constitute part of the Bonn Guidelines themselves,[56] but are addressed in the separate section C of the Decision VI/24.

activities within the CBD Secretariat For more details see the portal at: http://www.biodiv.org/programmes/socio-eco/traditional/default.asp.

[53] COP Dec VI/24, Part A, Section I, para 7.
[54] ibid para 1.
[55] ibid para 7b.
[56] COP Dec VI/24, Part A.

Some national access and benefit-sharing laws explicitly identify derivatives and require that benefits from their commercial use be paid. Perhaps the most renowned legal framework is the Andean Community Decision No 391,[57] which establishes a regime for common access to genetic resources. The access is rendered upon filing to the competent national authority of the access contract which contains all the requirements of rational use and environmental preservation without causing harm to the affected ecosystems in which the research and scientific activities are to take place.

Material Transfer Agreements (MTAs) are critical for the exchange of genetic resources. The Bonn Guidelines set the basic terms that should be included in a typical MTA. The CBD itself provides for access to be granted on Mutually Agreed Terms (MATs),[58] in other words users and providers must agree on certain terms in order to share the utilization and commercial use of genetic material. The critical elements that a standard MTA should include are:[59] reference to the CBD; legal status of the provider and the user; access and benefit-sharing provisions, including the permitted uses of genetic resources and their products or derivatives, the use of intellectual property rights, and the transfer to third parties; legal provisions that introduce the choice of law, dispute settlement and confidentiality arrangements.

Access to genetic resources is to be granted by the PIC.[60] Internationally this is conceived of as a legal principle embodied in the 1998 Rotterdam Convention on the Prior Informed Consent Procedure for Certain Hazardous Chemicals and Pesticides in International Trade.[61] While the PIC procedure, being principally linked to environmental law, refers to the duty to prevent harm and inform other countries of potential risks, its application in the access to genetic resources case under the CBD and the Bonn Guidelines is slightly different. It requires users of intermediaries who acquire genetic material to obtain consent from the owners of this genetic material. The Bonn Guidelines require that the PIC constitutes part of the entire access and benefit-sharing strategy and it extends responsibilities to both users and owners of genetic resources. The PIC procedure foreseen in the Guidelines should in general ensure legal clarity, cost-effectiveness, transparency, timeliness, and informed consent to all relevant stakeholders.

57 See Chapter 5 Section III.
58 CBD Art 15(4) and 15(7).
59 COP Dec VI/24, Part A, Appendix I.
60 ibid Section IV C.
61 Other agreements and codes of conduct employing PIC are the Convention on the Control of Transboundary Movements of Hazardous Wastes and Their Disposal (Basel, 22 Mar, 1989) (1989) 28 International Legal Materials, 649; the International Atomic Energy Agency (IAEA) Code of Practice on the International Transboundary Movement of Radioactive Waste, Resolution adopted by the General Conference, Vienna, Doc GC(XXXIV)/RES/530, 21 September 1990.

The Guidelines contain some standard elements that could be included in a typical PIC procedure. Finally, the Guidelines recognize the difficulties associated with obtaining access because of the diversity of stakeholders and their diverging interests, and therefore they acknowledge that their appropriate involvement in the system is to be judged on a case-by-case basis.

One practical step provided by the Guidelines is the establishment of focal points and competent authorities, responsible for providing information on procedures for gaining PIC and MATs as well as identification of the relevant stakeholders and competent national authorities through the CBD clearing-house mechanism.[62] Thus, national authorities shall either have the power to grant access to users, or choose to delegate such authority to other entities. They will also advise on matters regarding the negotiating process, and also requirements for PIC, national ABS arrangements, and mechanisms for the effective participation of stakeholders in the ABS process.

Unlike the CBD provisions, the benefit-sharing section of the Guidelines is quite comprehensive.[63] Its scope is broad as it covers 'all genetic resources and associated knowledge, innovations and practices covered by the CBD'.[64] The strategy set is composed of three principal elements: the identification of all the steps a user must follow to attain access and transparency of the system; establishment of a system for obtainment of PIC from the owner of the genetic resources; and the creation of MATs that are legally lucid, minimize costs and ensure that the providers' interests are met. However, the Guidelines do not form a multilateral system,[65] but they understand benefit-sharing as the balance among up-front, milestone payments and royalties on a case-by-case basis.[66] They also distinguish between monetary and non-monetary benefits.[67]

The delivery mechanism provided by the Guidelines is flexible, as partners come to arrangements suited to the circumstances based on mutually agreed terms,[68] an indicative list of which is included in the Guidelines.

### 2. *Positions of the parties during the negotiations of the Bonn Guidelines*

The negotiations of the Working Group on Access and Benefit-Sharing in Bonn, which elaborated the text of the Bonn Guidelines, were rather hard. The purpose here is to highlight the antagonizing interests that complicate and slow down negotiations. The experts' panel encountered resistance

[62] COP Dec VI/24, Part A Section II.

[63] ibid Section IV.

[64] ibid Section I, para 9.

[65] COP Dec VI/24, Part A Like the FAO International Treaty, see infra Chapter 4 Section 4.

[66] COP Dec VI/24, Part A Section IV para 47.

[67] ibid Section IV para 46 and Appendix II.

[68] ibid Section IV para 41.

from several parties and had to deal with numerous COP Decisions on access and benefit-sharing. The panel had to compile them and to come up with a politically acceptable text which could satisfy at least the majority of the parties. Issues of sovereignty, stakeholders, discrimination, derivatives and differentiation between user and provider obligations and intellectual property rights constituted highly debatable points.[69]

During the negotiations of the Bonn Guidelines and their adoption with Decision VI/24 at the COP 6, the European Communities (EC) underscored the need to harmonize the work and clarify the fields of competencies of the international fora and international instruments such as FAO and WIPO. It also stressed the need for a flexible and evolutionary text written in a balanced language that would not hinder national decision making. As regards the scope of the Guidelines, it insisted that the plant genetic resources for food and agriculture covered by the FAO IT should be explicitly excluded, but this language did not make the final text.

Developing countries that already dispose of access and benefit-sharing mechanisms were very sensitive to sovereign rights. They saw the debate over the Guidelines not so much about facilitating the development of regulatory frameworks for those lacking legislation, but as about the struggle between countries with access and benefit-sharing regulations already in place and those users accessing genetic resources. Developing countries also addressed proposals for the inclusion of information on the source of origin for genetic resources or evidence of prior informed consent for the use of traditional knowledge, deliberations which surfaced with the TRIPS Council.

The United States' (US) reaction was not as acute as it had been during the negotiating process of the International Treaty, evidently because the Guidelines are not binding and they require neither ratification nor endorsement. The Bonn Guidelines faced a lack of support from the US, particularly on intellectual property issues, because the Working Group's discussions on the matter were far more detailed than in most other CBD discussions. Thus, together with the biotechnology industry, the US highlighted potentially negative impacts and the burden that making such commitments obligatory would entail.

At the time, the international community was not yet ready for the adoption of a fully fledged treaty because of the various interests at stake. Even if there is a roadmap for 2010, it is debatable whether a fully fledged treaty will accommodate all the divergent interests, without vague and unclear stipulations.

[69] ENB, 'Sixth Meeting of the Conference of the Parties to the Convention on Biological Diversity (7–19 April 2002), Monday, 22 April 2002 (2002) 9 Earth's Negotiations Bulletin 239.

## IV. IMPLEMENTATION OPTIONS OF THE CBD ACCESS AND BENEFIT-SHARING PROVISIONS

### *A. Implementation Options in General*

The CBD's objectives could be implemented through the following mechanisms:

a) International instruments that promote the development of international binding norms to regulate access and benefit-sharing procedures. Following the WSSD call for an international access and benefit-sharing regime, negotiations have started in various international fora, such as the CBD, the FAO, the WIPO, and the WTO-TRIPS Council;
b) Regional agreements/national measures addressing access and benefit-sharing;
c) Private practices/professional society standards, such as codes of conduct and research guidelines.

Especially difficult is the implementation of the fairness and equity prerogatives of the CBD, for several reasons. First of all, concepts of fairness and equity are subjective and have different meanings in different places and at different points in time; unfortunately, the CBD does not provide a definition for what it considers being 'fair and equitable'.[70] In the elaboration of more precise rules on access and benefit-sharing, this uncertainty as to the interpretation of 'fairness and equity' impacts upon the public policy choices on property and intellectual property rights, and affects the final private or public law solutions.

Secondly, there are many stakeholders involved, including States, local communities, the private sector, and civil society, all of whom need to be considered in negotiating agreements. Thirdly, there is the problem of how to define/ evaluate the worth of genetic resources, which is not intrinsic to the resource itself and must therefore be decided by authorities; the final decision will necessarily affect the regulatory effort.

Once again, the final and heaviest burden of regulating access and benefit-sharing procedures falls on the state and national legislation.

At its Sixth Meeting in April 2002, the Conference of the Parties invited all parties and relevant organizations to submit information on existing measures on access and benefit-sharing to the Executive Secretary.[71] Although most parties did not respond to this call, the CBD Secretariat carried out research based on official sources and government websites. The

[70] See Chapter 2 Section III.B.3.
[71] COP Dec VI/24 D, para 6.

results were compiled into a database that, as of December 2008, included information on 27 countries.[72]

### *B. National Implementation Options*

Several theories have been developed for the implementation of international rules,[73] but independently of the theoretical background one embraces, it is clear that to become operational the CBD like most international rules need application within domestic legal systems. Several countries have passed laws on access to biological resources or are in the process of doing so. Most laws typically focus on research, collection, or exportation, but do not address the key issue of how to ensure the sharing of benefits derived from the use of genetic resources contained in biological specimens. The strategies that have emerged in provider countries with the aim of establishing access and benefit-sharing measures within their jurisdiction vary.

The first of these is the introduction of a single comprehensive law through either a parliamentary procedure or a presidential/legislative order in accordance with national constitutional rules.

A second strategy takes the form of a new set of laws for the implementation of a much broader set of objectives than the CBD, such as pursuance of general sustainable development principles, together with the development of special provisions on access and benefit-sharing.

A third strategy is to amend existing laws that were initially conceived for other purposes but that touch on access and benefit-sharing issues, such as the amendment of seed management laws to encompass the CBD requirements on PIC and MATs. The advantage of such an approach is that there are already present administrative structures, policies and institutions in support of the law, and it is cost effective, but the internal competences need to be redrawn.

In the absence of national legislation on this matter, individual bioprospecting contracts may be stipulated. The drawback of such contracts agreed upon between private persons/entities and local communities, individuals, or scientific institutions do not always guarantee the balanced and fair participation of all signatories.

[72] CBD Secretariat Database on ABS Measures at http://www.cbd.int/abs/measures.shtml (22/09/08).

[73] A Cassese, *International Law* (Oxford University Press, Oxford, 2005) 213 f.

### *C. Regional Implementation Options*

At the regional level, there are currently four agreements related to access and benefit-sharing. First there is the Andean Pact Decision 391 on the Common Regime on Access to Genetic Resources, which is a legally binding instrument and more elaborate than the other three regional instruments. Then there are the Central American Agreement on Access to Genetic Resources and Bio-Chemicals and related Traditional Knowledge and the ASEAN framework Agreement on Access to Biological and Genetic Resources, both in a draft form and not in force. Finally, the African Model Law for the Protection of the Rights of Local Communities, Farmers and Breeders, and for the Regulation of Access to Biological Resources provides a model for the African countries and it also addresses issues such as Farmers' Rights, plant breeders' rights and community rights and responsibilities.[74] It should also be mentioned that the Nordic region has adopted the Ministerial Declaration on Access and Rights to Genetic Resources in 2003, the Strategy for Conservation of Genetic Resources in the Nordic Region (2001–2004), and the Strategy for Genetic Resources in the Fisheries, Agriculture, Forestry and Food sectors in the Nordic Region (2005–2008).

Regional measures allow neighbouring provider countries with similar types of biological wealth to set the same access and benefit-sharing conditions to user countries and strengthen the regime promoted by the CBD.

## V. SUMMARY AND CONCLUSION

To sum up, the CBD itself does not lend detailed instructions on the implementation of its access and benefit-sharing provisions. It is clear however that access is essentially premised upon state sovereignty, mutually agreed terms and prior informed consent. Furthermore, the CBD provides for 'fair and equitable' benefit-sharing, but it does not give additional instructions on what the benefits should be and how they are to be distributed. Although indigenous peoples are directly affected by bioprospecting activities, the CBD does not set an obligation for respect of indigenous peoples' rights, but merely encourages such behavior. To address the challenges of CBD implementation, the COPs have extensively deliberated on the access and benefit-sharing issue and working groups have been established specif-

[74] CBD Note by the Executive Secretary, *Analysis of Existing National, Regional and International Legal Instruments relating to Access and Benefit-Sharing and Experience gained in their Implementation, including Identification of Gaps*, Ad Hoc Open-ended Group on Access an Benefit-Sharing, Third Meeting, Bangkok (14–18 February 2005), Doc UNEP/CBD/WG-ABS/3/2, 10 November 2004, para 76.

ically for dealing with issues of access and benefit-sharing and on CBD Article 8j. Finally, the Bonn Guidelines, a non binding instrument established with Decision VI/24, are a very elaborate tool supplying substantial help on access and benefit-sharing regulatory efforts, and a deadline has been set in 2010 for the elaboration and negotiation of a access and negotiation of an international regime. Despite this great advance with the Bonn Guidelines, the positions of the parties at the COPs coupled with the discordance again among the parties in the framework of access-and-benefit-sharing-related action at other international fora[75] demonstrate the difficulties for a single international instrument that aims to prescribe in detail access and benefit-sharing conditions.

Beyond the international attempts to regulate access and benefit-sharing, the implementation options portrayed are international instruments, national/regional implementation[76] and private initiatives.[77] It would not be precise to declare one alternative better that the other. Certainly, international supervision and co-ordination is needed, and the private practices' evolution is a very positive development from the users' performance perspective; however, source countries should count neither on an international treaty nor on the private practice, but they should enact their legislation and build domestic institutional capacity and, even better, they should try to engage in regional instruments.

[75] See Chapter 4. [76] See Chapter 5. [77] See Chapter 6.

# CHAPTER 4

# *The International Debate on the Emerging ABS Regime—FAO, WTO and WIPO*

## I. INTRODUCTION

As explained in the previous chapter on the implementation strategies for access and benefit-sharing, the promotion of an international treaty/protocol regulating access and benefit-sharing procedures has attracted much attention during the last decade. After the WSSD called for an international benefit-sharing regime, committees and expert groups have been established within the pertinent fora (CBD, WTO, WIPO) and negotiations are on the way for the realization of such an instrument. Since the CBD activities have been already dealt with, this chapter examines the work within the other organizations.[1]

The FAO activities are described with a view to drawing parallels between its multilateral system in effect for plant genetic resources, for food and agriculture, and the CBD bilateral system covering the rest of plant genetic resources. Further, the work within WTO-TRIPS Council and the WIPO is explored with an eye to understanding the intellectual property related facets of the access and benefit-sharing regime and to highlight the obstacles for the negotiation of a comprehensive single treaty on access and benefit-sharing. As a matter of fact, the WIPO has traditionally been the international forum for the harmonization of global intellectual property rules, but the inclusion of the TRIPS Agreement in the WTO system presses states to account for the intellectual property matters within the TRIPS Council as well.

## II. THE CALL FOR AN INTERNATIONAL REGIME AT THE WSSD, JOHANNESBURG 2002

The World Summit on Sustainable Development was held in Johannesburg

[1] As of December 2008. A previous version of the following section appeared at A Smagadi, 'The Impact of the Convention on Biological Diversity on the Utilization of Plant Genetic Resources and Benefit-Sharing' (2006) 6 Yearbook of European Environmental Law 120.

from 26 August to 4 September 2002. Biodiversity was recognized as playing a critical role in overall efforts to achieve sustainable development and poverty eradication. It was one of the five key areas in the United Nations 'WEHAB' initiative, along with water, energy, health and agriculture.[2] The Plan of Implementation, specifically refers to the Bonn Guidelines in its paragraph 44(o): 'within the framework of the Convention on Biological Diversity, bearing in mind the Bonn Guidelines, an international regime to promote and safeguard the fair and equitable sharing of benefits arising out of the utilization of genetic resources'.

## III. INTERNATIONAL INITIATIVES FOR THE PROTECTION OF TRADITIONAL KNOWLEDGE

Traditional knowledge has extensively been discussed in numerous books and papers. In the production of medicines, traditional knowledge can be a fundamental element of biomedical research. Genetic resources alone are not as useful as if they are accompanied by the knowledge and experience of the local communities, whose individuals may guide bioprospectors concerning the qualities of plant genetic material and its potential uses. The efforts of researchers are thus automatically diminished. Further research and examination may lead to superior innovative pharmaceutical products. Unlike the easy-to-define 'genetic resources' term, 'traditional knowledge' has a number of different subsets, some of them designated by expressions such as 'indigenous knowledge', 'folklore', or 'traditional medicinal knowledge'.[3]

The notion of traditional knowledge is commonly perceived as being related to ancient knowledge. However, this is not true; traditional knowledge has the intrinsic characteristic of being evolutionary and dynamic, it is developing all the time in a process of continuous creation as individuals and communities absorb the challenges of their physical and cultural environment. The commercial and other advantages deriving from its use are multiplied by international trade, communications, and cultural exchange; and it gives rise to intellectual property questions. The sudden run after this knowledge by global traders has made its custodians conscious of the economic potential of the information they possess.[4]

[2] United Nations Report of the World Summit on Sustainable Development, Johannesburg (26 Aug–4 Sept 2002) Doc A/CONF.199/20, United Nations, New York.

[3] See also lso WIPO–IGC, *Traditional Knowledge—Operational Terms and Definitions*, 3rd Session, Doc WIPO/GRTKF/IC/3/9, 20 May 2002.

[4] See JL Trotti, 'Compensation Versus Colonization: A Common Heritage Approach to the Use of Indigenous Medicine in Developing Western Pharmaceuticals' (2001) 56 Food and Drug Law Journal 367; DR Downes, 'How Intellectual Property Could Be a Tool to Protect Traditional Knowledge' (2000) 25 Columbia Journal of Environmental Law 253; N Roht-Arriaza, 'Of Seeds and Shamans: the Appropriation of the Scientific and Technical Knowledge of Indigenous and

A major problem identified in analysing traditional knowledge and cultural expressions is that indigenous peoples do not consider their heritage in terms of property at all, but in terms of community and individual responsibility. After all, some medicinal practices have sacred significance and their commercial use contradicts the beliefs and the cultural values of the indigenous population. The inadequacy of the present Western intellectual property system, and the difficulties brought about by its imposition on developing countries and local communities suggests a *sui generis* solution not necessarily implying commoditization.[5]

Initial efforts for the protection of traditional knowledge were made in the 1980s by a campaign conducted jointly by the United Nations Educational, Scientific and Cultural Organization (UNESCO) and the WIPO,[6] with the objective to set forth a set of norms for the protection of expressions of folklore against illicit exploitation. The outcome of this campaign, the Model Provisions For National Laws On The Protection Of Expressions Of Folklore Against Illicit Exploitation and Other Prejudicial Actions UNESCO & WIPO in 1985, enunciated general principles, such as the contribution of communities' knowledge and the need for its development, but the identification of this knowledge and the methods of exploitation were left to the national governments. The UNESCO/WIPO document limits the definition of folklore to verbal, musical, and tangible expressions, and expressions by action, leaving aside folk medicines that are also part of the folklore community. But its failure was more due to the lack of political will to implement it, than its weakness in content.

The International Labour Organization Convention 169 (ILO C 169), concerning Indigenous and Tribal Peoples in Independent Countries, entered into force on 5 September 1991, and was to some extent more successful in terms of indigenous people's rights than the UNESCO/WIPO Model Law. While it calls attention to the distinctive contribution of indigenous peoples to cultural diversity and the respect and recognition of their

Local Communities' (1996) 17 Michigan Journal of International Law 919; United Nations Conference on Trade and Development, *Protecting and Promoting Traditional Knowledge: Systems, National Experiences and International Dimensions* (UN Publications, New York and Geneva, 2004); M Halewood, 'Indigenous and Local Knowledge in International Law: A Preface to Sui Generis Intellectual Property Protection'(1999) 44 McGill Law Journal 953.

[5] NS Gopalakrishnan, 'Protection of Traditional Knowledge—The Need for a Sui Generis Law in India' (2002) 5 The Journal of World Intellectual Property 740; M Blakeney, 'Bioprospecting and the Protection of Traditional Knowledge of Indigenous Peoples: An Australian Perspective'(1997) 6 European Intellectual Property Review 300; Y Liu, 'IPR Protection for New Traditional Knowledge: With a Case Study of Traditional Chinese Medicine' (2003) 25 European Intellectual Property Review 194; E Da Silva, 'The Protection of Intellectual Property for Local Indigenous Communities' (1995) 11 European Intellectual Property Review 546.

[6] World Intellectual Property Organization, *Intellectual Property Needs and Expectations of Traditional Knowledge* (WIPO, 2001) 93.

social and cultural values, ILO C 169 does not provide a solid basis for the protection of indigenous knowledge.

Finally, the UN Declaration on the Rights of Indigenous People[7] expresses the need to recognize the full ownership, control and protection of peoples' cultural and intellectual property. This includes the right to special measures to control, develop and protect their sciences, technologies, cultural manifestations, seeds, medicines, knowledge of the properties of flora and fauna, oral traditions, literature, design and visual and performing arts (Article 29). While the language of the Declaration is more determined than the previous international documents on the protection of traditional knowledge, it does not generate binding obligations for the states. Some states have indeed instituted legislation for the protection of indigenous peoples and one of the most renowned examples is the Philippines legislation.[8]

The CBD, although confined to the issue of the traditional knowledge over genetic resources, lays down important norms for the protection of this knowledge. While recognizing the sovereign rights of nations over their natural resources (Article 15.1), the CBD mandates the acknowledgement of holders' rights to the knowledge in case of their use. According to Article 8j CBD, the obligations of the parties consist of:

- the respect, preservation and maintenance of the knowledge, innovations and practices of indigenous and local communities;
- the promotion of its wider application with the approval and involvement of the holders of such knowledge, innovations and practices, and;
- the encouragement of equitable benefit-sharing from the utilization of such knowledge, innovations, and practices.

The CBD mandate recognizes the ownership rights of the community and therefore it dictates the relevant community's involvement and approval before the use of its knowledge and the share of the benefits. Individual nations have to follow this obligation, but they are free to shape the mechanism by which it can be fulfilled. At this point the usual benefit-sharing problems appear, about the legal, administrative and institutional capacities of developing countries to meet such an obligation; the insistence of industrialized countries to enforce their 'good practice' regulatory frameworks; and the claims of the various stakeholders.

The difficulties are manifold and an additional impediment is that in many developing countries the relationship of the government with the indigenous

[7] United Nations Declaration on the Rights of Indigenous Peoples, Doc. A/RES/61/295 adopted by the General Assembly on 13 September 2007.

[8] See Chapter 5 Section II.

local communities is almost non-existent or it is delineated by particular rules completely outside Western societies' organizational schemes.

Within the CBD Secretariat there is the Working Group on the Implementation of Article 8(j) and Related Provisions. The Secretariat also cooperates with other UN agencies, such as the FAO, the WIPO, the WTO, the UN Forum on Forests, the Commission on Human Rights Working Group on Indigenous Populations, and the UN Conference on Trade and Development (UNCTAD). This collaboration aims at ensuring that issues concerning the protection and application of traditional knowledge, innovations and practices, and the involvement of indigenous and local communities in biodiversity-related activities, are given the widest possible focus.

Through the years, the challenges for the protection of traditional knowledge have intensified, so has the number of international bodies working on the issue. Apart from the CBD Conferences of the Parties and the subsidiary bodies that tackle traditional knowledge in the context of a more general inquiry, that of an inclusive regime on access and benefit-sharing, WIPO and WTO have developed activities in this regard. The increasing demand to expressly recognize and protect traditional knowledge under TRIPS—in the form of patents, geographical indications, traditionally undisclosed information as well—led to the WTO Ministerial Declaration in Doha instructing the TRIPS Council to examine and report on the interrelationship between the TRIPS, the CBD and the protection of traditional knowledge and folklore.[9] Also the absence of uniform norms led to the setting up of an Intergovernmental Committee on Intellectual Property and Genetic Resources, Traditional Knowledge and Folklore by the WIPO.[10] WIPO and WTO will be examined more extensively, after a review of the work undertaken by the FAO.

## IV. THE FOOD AND AGRICULTURE ORGANIZATION

Together with the FAO IT, the Bonn Guidelines form the international body of ABS regulation. The FAO international rules developed from a number of concerns separate from, but related to, the CBD process. The two instruments share the same goals, but represent different approaches. The rules set out in the IT relate exclusively to the domain of plant genetic resources for food and agriculture, and it is interesting to draw the parallels between the IT and the Guidelines.

[9] See Section V. [10] See Section VI.

### *A. From the International Undertaking to the International Treaty*

In the 1980's higher-value export commodities or genetically modified varieties phased out most traditional crops. This development occurred very rapidly and raised great concerns about food security, because it seed supply was no longer allowed and farmers' survival was threatened. At the same time, rapid global population growth called for greater food production.

In response, at the FAO General Conference in 1983, the Commission on Genetic Resources for Food and Agriculture (CGRFA) was established and the International Undertaking was adopted as a non-binding resolution, to 'ensure that plant genetic resources of economic and/or social interest, particularly for agriculture, will be explored, preserved, evaluated and made available for plant breeding and scientific purposes'.[11]

The Commission's work focused on a rights-based approach to the farmers' contribution in the overall production and in 1989, farmers' rights were officially introduced for the first time in the FAO Resolution 5/89 that recognized the historical role of farmers in the improvement of crops and the need to ensure their access to seeds

Later, in 1991, after seven years of negotiations, the International Treaty of Plant Genetic Resources for Food and Agriculture was adopted, reflecting the new orientation given by the CBD: the IT emphasizes the conservation of plant genetic resources for food and agriculture, their sustainable use, and their benefit-sharing, with the ultimate goal to promote sustainable agriculture and food security. It is the first treaty providing a legal framework that delineates a regime for access and benefit-sharing; it links plant genetic resource conservation, intellectual property rights, sustainable agriculture, and food security, marking a novel approach in international law.[12]

While the IT is distinguished for recognizing farmers' rights, which are not addressed by any other international treaty, it does not directly handle patents or plant breeders' rights,[13] which are covered by the TRIPS and the Convention for the Protection of New Varieties of Plant (UPOV Convention)[14] respectively.

The IT entered into force on June 29, 2004, after numerous appeals for its ratification in due time, including the Declaration of the World Food Summit, the Ministerial Declaration of the COP6, and the Plan of Implementation of the Johannesburg World Summit on Sustainable Development.

[11] IU Article 1.

[12] IT Art 12.3f.

[13] IT Part III and in particular Art 9 and Art 13(3).

[14] UPOV stands for the French Union Internationale pour la protection des obtentions végétales.

### *B. The Concept of Farmers' Rights*

Since it was first enshrined in the IU, the concept of farmers' rights has been contentious. It does not include property rights, but presupposes that States develop some form of protection. The term should thus be clearly distinguished from 'plant breeders' rights' and 'farmers' privilege'. Plant breeders' rights are a form of intellectual property rights specifically designed to protect plant varieties; the provisions of the UPOV come into play for the protection of these rights. Related to plant breeders' rights, the farmers' privilege limits the scope of plant breeders' rights by allowing farmers to replant propagating material of plant varieties protected by plant breeders' rights, if this material was harvested on their own holdings.[15]

Overall the concept of farmers' rights is interesting because it fosters a broader understanding of the links between innovation, rights over knowledge, biodiversity conservation and the sustainable use of agro-biodiversity, whereas patents and breeders' rights are separated from concerns over conservation and sustainable use. The IT acknowledges the farming communities' contribution to the development and maintenance of agricultural biodiversity,[16] but in practical terms, this acknowledgement only contributes to the progressive development of the notion of benefit-sharing, but is far from an international policy decision for actual compensation.

### *C. Comparison of the Benefit-Sharing Legislation under the Bonn Guidelines and the International Treaty on Plant Genetic Resources for Food and Agriculture*[17]

The IT is intended to be a legally binding agreement under international law. The implementation of the Bonn Guidelines depends on the good will of the parties to the CBD. However, as already stressed above, there is a hope that the Bonn Guidelines represent the first step towards the formulation of a treaty very soon.

Unlike the Bonn Guidelines, which use the bilateral approach to benefit-sharing, the IT covers all genetic material for food and agriculture by specifically putting in place a multilateral system to facilitate access to 64 essential crops and their varieties. Parties deal directly with the system and not with each other; and benefit-sharing arrangements are possible only in the context of the genetic resources listed in the multilateral system. At this

[15] MA Girsberger, 'The Concept of Farmers' Rights, the Role of FAO and the TRIPS Agreement' in TCottier and PC Mavroidis (eds) *Intellectual Property—Trade, Competition and Development* (University of Michigan Press, Ann Arbor, 2003) 455–479.

[16] IT Art 9.1.

[17] See also this discussion at WB Chambers, 'Emerging International Rules on the Commercialization of Genetic Resources—The FAO International Plant Genetic Treaty and CBD Bonn Guidelines' (2003) 6 The Journal of World Intellectual Property 314.

point, the system envisaged by the Bonn Guidelines is broader in scope since it covers all biodiversity. The multilateral approach was selected by the IT, in consideration of the enormous transaction costs arising from bilateral agreements and the difficulties to coordinate.[18] In terms of type of benefit-sharing arrangements, the IT defines the exchange of information, access to transfer of technology, and capacity-building (Art. 13.2) as benefit-sharing mechanisms, whereas, the Bonn Guidelines are not that specific and deal with benefit-sharing on a case-by-case basis.

Because of its multilateral approach, the IT sets up an institutional mechanism to oversee its implementation, the Governing Body. Of particular importance is the role of the Governing Body in the delivery mechanism foreseen in the IT: unlike the Bonn Guidelines, where distribution of the benefits is agreed by the parties depending on the specific circumstances, the benefits in the IT regime are to be broadly shared with farmers in all countries and with States themselves. Prominence is given to farmers in developing countries and countries with economies in transition.[19] The benefits are to be delivered by means of a common fund administered by the IT's Governing Body. The Fund calls for those parties that use material accessed from the multilateral system and create commercial value from the resource to pay an 'equitable share' of the benefits. What constitutes 'equitable share' is however not clearly defined.

The Bonn Guidelines leave open a number of controversial areas, such as the issue of benefits arising out of derivatives. Similarly, the IT does not clarify whether benefits from derivatives are to be contributed to the established fund and administered by the Governing Body of the Treaty.

Also, the question of intellectual property rights on the genetic parts of the crops listed in the multilateral system needs further clarification. In fact, all issues surrounding intellectual property are highly contentious, because of the stakes involved; the discussion on intellectual property rights and whether components of a genetic resource are patentable was one of the main points of controversy between the US and developing countries— which had the tacit support of the EU— in the IT negotiations and eventually led to the controversial article, 12.3(d). Further to Article 12.4, the debate on Article 12.3(d) will be solved by the Governing Body, implying some sort of political influences in its interpretation. On intellectual property, the Bonn Guidelines do not set forth any specific restrictions; intellectual property rights are handled in a separate section of the same Decision VI/24 that includes the Bonn Guidelines.

[18] See presentation of the FAO representative at the TRIPS Council, WTO–TRIPS Council, *Minutes of Meeting—Held in the Centre William Rappard on 25–27 and 29 November, and 20 December* 2002, Doc IP/C/M/38, 5 February 2003, paras 219 f.

[19] IT Art 13(3).

On the MTAs, the Bonn Guidelines include the essential elements that should be included in a standard MTA. The IT lays down some basic conditions,[20] and in 2006, the IT Governing Body at its first session adopted the Standard Material Transfer Agreement.[21]

Another difference between the Bonn Guidelines and the IT is the PIC. In the Bonn Guidelines, PIC is granted by the contracting party providing the source material, whereas in the IT, PIC is granted by the central authority. The IT recognizes the sovereign rights of the countries over their genetic resources, but these countries give up their right of PIC and bestow this responsibility on the IT. The IT will in turn ensure the protection and use of plant genetic resources and benefit-sharing for the common good of the parties.[22] This was the result of the acknowledgement that there was much difficulty in determining the country of origin of the genetic material, as many of the resources in the Multilateral System are widely found *ex situ* in a wide array of countries.

In effect, the degree to which the international rules set by the agreements are realized is directly linked to the capacity that can be provided at the national level to implement them. This is closely related to capacity building programs for the planning of national implementing strategies. Issues of both instruments that still need to be addressed involve the further elaboration of the MAT in the IT, the use of terms particularly in the Bonn Guidelines and certainly the issue of derivatives.

## V. THE WORLD TRADE ORGANIZATION—TRIPS

### A. *Short Historical Overview: TRIPS Agreement*

The Uruguay Round introduced, for the first time since the GATT was drafted in 1947, multilateral negotiations on 'trade-related intellectual property rights'.[23] Industrialized countries insisted upon including intellectual property rights in the Uruguay negotiations. US policy-makers in particular exerted pressure for the establishment of a multilateral intellectual property system as a result of intense lobbying by the US pharmaceutical industry. The latter insisted on the linkages between intellectual property and international trade and succeeded in making intellectual prop-

[20] See IT Art 12(4) and 12(3)(a), (d) and (g).

[21] Resolution 2/2006, see Food and Agriculture Organization, *First Session of the Governing Body of the International Treaty on Plant Genetic Resources for Food and Agriculture*, Madrid, Spain (12–16 June 2006), Doc IT/GB-1/06/Report.

[22] IT Preamble and Art 10.

[23] See F-K Beier and G Schricker (eds) *From GATT to TRIPs: The Agreement on Trade-related Aspects of Intellectual Property Rights* (New York, Weinheim, 1996).

erty protection a central aim of US foreign economic policy.[24] Having the support of other industrialized countries, such as Canada, Japan and the EU, the US argued that intellectual property protection should be considered a trade issue because of the growing internationalization and technology exchange.

In addition to the TRIPS international regime, the US makes use of its domestic intellectual property laws to put pressure on less developed countries. Before the existence of the TRIPS agreement, the lack of minimum international standards and an inadequate international enforcement mechanism led to the use of unilateral action. Section 182 of the Omnibus Trade Act of 1974, commonly referred to as 'special 301', has placed great pressure on developing countries to meet the US demands. The US Trade Representative must name 'priority countries', namely countries of considerable economic interest for the US, which do not negotiate in good faith or make serious progress in negotiations initiated to improve their intellectual property laws. After the identification of these countries, an investigation is carried out that may lead to retaliation in the form of trade sanctions.[25]

These attempts to formulate intellectual property as a trade issue have irritated many developing countries who claim that intellectual property is part of the national legal system and GATT is not the appropriate forum to put pressure on countries over the content of their domestic legislations. They believed that intellectual property at the international level is to be addressed by WIPO, which was established exactly for this purpose. The WIPO, however, does not have the enforcement mechanism that the WTO has, and can therefore only make recommendations.

India, for example, a developing country with a strong pharmaceutical industry, has resisted the enforcement of the Western type intellectual property regime and relies on the manufacture of generic drugs. Industrialized countries combat generic drugs production because this competes with the interests of their pharmaceutical companies that produce the same drugs and charge higher prices. In any event, local needs shape the course for countries in choosing policy options, and here not only India, but also Brazil and South Africa present distinctive cases. Their main argument is that the poverty of their populations and their health emergencies, such as AIDS and other epidemics, must take precedence over compliance with the

[24] CM Correa, *Intellectual Property Rights, the WTO, and Developing Countries: the TRIPS Agreement and Policy Options* (Zed Books, New York, 1999) 5 and 23 f; CM Correa and AA Yusuf (eds) *Intellectual Property and International Trade: the TRIPS Agreement* (Kluwer Law International, London, Boston, 1998) 4 f; E Henderson, 'TRIPS and the Third World: the Example of Pharmaceutical Patents in India' (1997) 19 European Intellectual Property Review 651.

[25] Henderson ibid 652.

international intellectual property norms. Global welfare can never be the excuse for local distress.[26]

While such an argument is not easily accepted by the industrialized world, it should be noted that Canada and the US completely changed their policy during the Anthrax crisis and the threat of bio-terrorism that started in 2001.[27] They both allowed for the compulsory licensing and the generic production of medicines in order to confront the national emergency, although they do not easily accept the same policy measures when developing countries are facing national emergencies.

On this particular matter relating to access to medicines, the manufacture of generic drugs, and the developing countries' claims to the right to the health of their populations, the Doha development Agenda[28] has given the mandate for a special series of discussions within the TRIPS council. It is one of the hottest issues in the international trade agenda, and it highlights one of the deficiencies of the first years of the TRIPS implementation. Together with the issue of potential controversies between the TRIPS and the CBD, and the geographical indications, the problem of access to medicines and parallel imports is one more example of the tensions between developed and developing countries in the enforcement of a global intellectual property regime.

### *B. Key Provisions on Patentable Inventions*

The Trade Related Aspects of Intellectual Property Rights Agreement (TRIPS) establishes precisely the kind of international minimum standards for adequate and effective intellectual property protection, including standards for the compulsory licensing of patented technology. It sets up minimum standards on copyright and related rights (including computer products and databases), trademarks, geographical indications, industrial designs, patents, integrated circuits, and undisclosed information (trade secrets).[29]

Articles 1–8 of TRIPS provide for the scope of application of national treatment, most-favored nation, and exhaustion of rights. It requires that members comply with the substantive provision of the Paris and Bern

[26] There are numerous studies on the welfare effects of the introduction of strong intellectual property rights in developing countries, see Correa (n 24) 25.

[27] S Ragavan, 'Can't We All Get Along?—The Case for a Workable Patent Mode' (2003) 35 Arizona State Law Journal 117.

[28] Clause 4 of the Declaration and the separate Declaration on Public Health.

[29] For a compact description of the TRIPS Agreement patent provisions see DJ Gervais, 'The TRIPS Agreement—Interpretation and Implementation' (1999) 21 European Intellectual Property review 156; J Straus, 'Bedeutung des TRIPS für das Patentrecht' (1996) Heft 3 Gewerblicher Rechtschutz und Urheberrecht Internationaler Tei, 179; See also Correa and Yusuf (n 24) 12 f.

Conventions[30] administered by the WIPO. Thus these conventions become binding for WTO Members who are not parties to them and the disputes arising are now subject to the notable WTO dispute settlement procedure. Of particular importance is Article 6 which excludes from the dispute settlement the question of 'exhaustion of rights', otherwise known as parallel importation. That is the importation of goods lawfully manufactured in another country and generally intended for distribution in that other country.

The patent criteria are set out in Articles 27 through 34 of TRIPS in a comprehensive and flexible manner.[31] The patent section of the Agreement was the most difficult to negotiate, because it was related to a wide range of North-South issues.

Article 27.1 requires that patents be available in all fields of technology, based on the usual three criteria of novelty, inventive step, and industrial application. By stating that products and processes are subject to patentability and that no discrimination is to be made on the field of technology, the TRIPS introduces a general principle of their eligibility to be patented. The TRIPS also eliminates discrimination according to the place of invention and whether the product is imported or produced locally.

A serious restriction to the principle of eligibility to be patented is set forth in Article 27.2. The article stipulates that a WTO member may exclude inventions from patentability based on a risk that their commercial exploitation within its territory could endanger *ordre public* or morality. Examples given are the protection of human, animal or plant life or health or serious prejudice to the environment. How broadly these provisions may be interpreted is not clear. Concepts such as '*ordre public*', 'morality' or 'serious prejudice' are not precise. Therefore, it has been suggested that an objective justification be included, whereby the exclusions of specific inventions rather than of entire categories of inventions are excluded from patentability.[32]

In Article 31 reference is made to compulsory licenses. Specific conditions are provided for the granting of compulsory licenses, and the article constitutes a detailed check-list of WTO Members. Finally, Article 33 establishes a minimum term of protection of 20 years from filing.

### C. *Legal Tensions between TRIPS and the CBD*

Conflicts arose from the formulation of Article 27.3(b), which reads:

[30] TRIPS Art 2.
[31] Correa and Yusuf (eds) (1998 n 24) 189 f.
[32] Gervais (1999 n 29) 159.

> 3. Members may also exclude from patentability
> (a)...
> (b) plants and animals other than micro-organisms, and essentially biological processes for the production of plants or animals other than non-biological and microbiological processes. However, Members shall provide for the protection of plant varieties either by patents or by an effective sui generis system or by any combination thereof. The provisions of this subparagraph shall be reviewed four years after the date of entry into force of the WTO Agreement.

The review of Article 27.3(b) is closely related to the relationship between the CBD and the TRIPS Agreement and to the protection of traditional knowledge.[33] The main problem raised by the parties is the broad provisions of patent protection for plant animal inventions that allow for the patenting of life forms. This raises a wide range of food security, environmental, cultural and moral concerns. Developing countries claim that the current international intellectual property system allows for: granting of excessively broad patents and brings about biopiracy of genetic resources and associated traditional knowledge; that the modification and displacement of seeds and traditional plant varieties cause biodiversity depletion; and that there is a need to rebalance with the CBD, in other words a need to integrate the CBD principles in the TRIPS Agreement, so that they become effective. Industrialized countries, on the contrary, support that the actual state of international intellectual property rules is more than adequate for the promotion of private sector investment and facilitation of technology transfer, and deny any kind of conflict between the two agreements.[34]

In the context of the current thesis, issues surrounding Genetically Modified Organisms (GMOs) (see patentability exceptions of Article 27.3b and 27.2) and agricultural concerns (effective *sui generis* protection of plant varieties) will be left aside. Focus will be placed on efforts to avoid

[33] See the three complementary notes of the WTO Secretariat: WTO–TRIPS Council, *The Relationship between the TRIPS Agreement and the Convention on Biological Diversity—Summary of Issues Raised and Points Made*, Doc IP/C/W/368/ Rev.1, 8 February 2006 and WTO Secretariat: WTO–TRIPS Council, *The Relationship between the TRIPS Agreement and the Convention on Biological Diversity—Summary of Issues Raised and Points Made*, Doc IP/C/W/368/ Rev.1/Corr.1, 9 March 2006; WTO–TRIPS Council, *Review of the Provisions of Article 27.3(b)—Summary of Issues Raised and Points Made*, Doc IP/C/W/ Rev.1, 9 March 2006; and WTO–TRIPS Council, *The Protection of Traditional Knowledge and Folklore—Summary of Issues Raised and Points Made*, Doc IP/C/W/370/Rev.1, 9 March 2006; the three documents constitute a revision of the initial documents prepared by the Secretariat on 8 August 2002. Their annexes comprise a detailed list of all relevant documents submitted to the Council by the members, organizations and secretariat notes on the three issues.

[34] ibid.

biopiracy and on calls for making the CBD principles operational through the TRIPS Agreement.

### *D. Doha Development Agenda and Beyond*

In Doha, on 14 November 2001 (five years after the entry into force of the WTO Agreement), the Ministers decided to launch negotiations for the review of Article 27.3(b) TRIPS.[35] According to the Doha Declaration, the TRIPS Council must take into consideration the following in its review work or in dealing with any other implementation issue: the relationship between the TRIPS Agreement and the UN Convention on Biological Diversity, the protection of traditional knowledge and folklore, and other relevant new developments raised by member governments. The objectives and principles of the TRIPS Agreement (Articles 7 and 8 respectively), in addition to issues concerning development are also to be taken fully into account.

Since then, the TRIPS Council has started to work on the three issues; not only were WTO members requested to contribute to the TRIPS work with their own considerations and proposals, but other international organizations were also asked to participate.[36] Among these were: the CBD Secretariat, the Consultative Group on International Agricultural Research (CGIAR), the FAO, the UNCTAD, the United Nations Environment Programme (UNEP), the UPOV, the WIPO and the World Bank. As part of their contribution, they were requested to provide updated information on their activities affecting the three interrelated issues: the relationship between the TRIPS and the CBD, the review of the provisions of Art. 27.3(b), and the protection of traditional knowledge and folklore.

It should be noted, however, that since the Doha mandate was launched in November 2001, little work has been done in the framework of the TRIPS Council, not only because the Council is preoccupied with public health issues and access to medicines;[37] but also because of the political relevance of the negotiations at the WIPO Intergovernmental Committee for the modification of intellectual property treaties and elaboration of a new intellectual property tool for the protection of traditional knowledge and at the CBD Conferences of the Parties for the development of an access and benefit-sharing instrument after the call of the Johannesburg summit. The December 2005 Hong Kong Ministerial Conference repeated the Doha mandate to the TRIPS Council.[38]

[35] Doha Ministerial Declaration, 14 November 2001 para 19.

[36] Doc IP/C/M/38, para 217.

[37] Doha Declaration para 17; and Declaration on the TRIPS Agreement and Public Health.

[38] Hong Kong Declaration para 44. The Draft Declaration circulated during the Cancún Ministerial in September 2003 comprised the identical paragraph 23.

The discussions are still ongoing and there is no definitive solution in sight, because of the diversity of solutions suggested and also because the parties disagree about the nature and essence of the problem itself, as will be shown in the following sections.[39] In sum, the general issue raised concerns whether there is any conflict between the TRIPS and the CBD, and if there is, whether it warrants a review of the TRIPS. Should the PIC and benefit-sharing provisions be included in the TRIPS? Do the patentability criteria set by the TRIPS Agreement impede the materialization of the CBD's objectives for sustainable management and benefit-sharing? Does the inadequate protection of traditional knowledge associated to genetic resources and used for the production of medicines reduce the incentives for preservation and sustainable use of biodiversity by the indigenous populations and hence limit the implementation of the CBD? The different views of the Member States in TRIPS Council are briefly presented in the following section.

### *E. Positions of the Parties on CBD–TRIPS Issue*

Obviously the positions reflect the different economic interests of the parties. There are four broad categories of the views expressed with regards to the existence of a conflict between the CBD and the TRIPS and to the need of any action within TRIPS, such as amendment of the TRIPS Articles 27.3(b) and 29 to foster a balanced and mutually supportive implementation of both. Whether there is a conflict between the CBD and TRIPS is also treated within the WTO Committee on Trade and Environment (CTE) in the context of the discussions of conflicts appearing between the WTO and the Multilateral Environmental Agreements (MEAs).[40]

The four views can be summarized as follows:[41]

a) There is no conflict between TRIPS and the CBD and the governments can implement both in a mutually supportive way by taking the necessary national measures (combined contracts and domestic measures).
b) There is no conflict, but while countries take the necessary domestic measures to implement them, there is a need to further examine whether concurrent international action regarding patents is necessary.

[39] See also the recent WTO—General Council, Trade Negotiations Committee, *Doha Work Programme—the outstanding Implementation Issue on the Relationship between the TRIPS Agreement and the Convention on Biological Diversity—Communication from Brazil, India, Pakistan, Peru, Thailand and Tanzania*, Doc. WT/GC/W/564 and TN/C/W/41, 31 May 2006.

[40] Decision on Trade and Environment, adopted by the ministers at the meeting of the Uruguay Round Trade Negotiations Committee in Marrakesh on 14 April 1994.

[41] See Doc IP/C/W/368/ Rev.1, paras 6 f.

c) There is actually an inherent conflict and it is indispensable to amend the TRIPS Agreement to resolve it (TRIPS amendment to include obligation for disclosure).
d) There is no inherent conflict, but potential conflict could arise from the implementation of the two agreements, and here international action is necessary to safeguard the mutually supporting functions of the TRIPS and the CBD.

The first position of national-based approach is primarily supported by the US,[42] followed by Japan, Korea and Singapore. According to this view, the TRIPS and the CBD have different objectives and subject-matter, and there is thus no specific case for conflict. The process of granting patent rights does not prevent compliance with the CBD principles of State sovereignty, prior informed consent and benefit-sharing. WTO Members who are at the same time contracting parties to the CBD can comply with their obligations under both agreements with an emphasis on the complementarity of the two through the use of an access regime for genetic resources on the basis of a contracts coupled with domestic measures.

Further to this position, contracts authorizing collection of genetic material should include provisions requiring reporting and benefit-sharing, and the obligation of the user to notify the provider on any invention potentially developed from the material collected under the contract. Domestic legislative and administrative measures should further regulate on the access and benefit-sharing contracts.

As for the erroneously granted patents, the proponents of this view claim that the current patent system provides for remedial proceedings, such as re-examination and revocation. Cooperation and dissemination of information would consolidate the existing system with regards to what constitutes 'prior art'.[43]

The second position, closely related to the first one above, is promoted by Australia, Canada and New Zealand. Similarly, these countries deny any conflict between CBD and TRIPS and deem any proposed amendment of the TRIPS in this matter unnecessary. They acknowledge the need to prevent biopiracy through domestic measures supported through international action. Such action would include information sharing and databases, short of TRIPS amendment. They propound a 'fact-based' discussion

[42] A number of communications have been submitted by the US, the most recent in 2006, see WTO–TRIPS, *Article 27.3(b), Relationship between the TRIPS Agreement and the CBD, and the Protection of Traditional Knowledge and Folklore—Communication from the United States*, Doc IP/C/W/469, 13 March 2006.

[43] See WTO–TRIPS Council, *Technology Transfer Practices of the US National Cancer Institute's Departmental Therapeutic Programme—Communication from the United States*, Doc IP/C/W/341, 25 March 2002, and WTO–TRIPS Council, *Access to Genetic Resources Regime of the United States National Parks*, Doc IP/C/W/393, 28 January 2003.

on the matter, in other words a thorough review and analysis of access and benefit-sharing legislation and related implementation to pinpoint the failures and address them.

The third position, primarily taken by the group of developing countries, fully opposed to the views of most industrialized countries, requests the amendment of the patent system to clearly integrate the CBD requirements, to overcome the conflict between CBD and TRIPS.[44] This conflict between the two agreement lies in the fact that TRIPS provides for the appropriation of genetic material by private parties, whereas the CBD stresses the countries' sovereign rights over genetic material. At the same time TRIPS oversees protection of traditional knowledge.

According to this position, the cross-border character of genetic resources, the intricate character of traditional knowledge and the asymmetry in the bargaining powers of the parties to any access and benefit-sharing contract cannot be addressed by a fragmented and costly nation-by-nation system. For this reason, an international solution should be accepted as a common denominator that would incorporate the CBD requirement in TRIPS through amendment. Such an action, would favor developing countries that lack administrative and financial capacity to follow biopiracy cases.

In 2004, the States supporting this view specified their arguments with a checklist of issues to be discussed.[45] Accordingly, the international intellectual property regime should be amended to condition the granting of a patent upon disclosure of the source or country of origin of the biological resource and the traditional knowledge used in the invention; evidence of prior informed consent through the approval of authorities under relevant national regimes; and evidence of a fair and equitable share of benefits. This

[44] A number of communications have been submitted either separately by the countries, or jointly by the group, see eg WTO–TRIPS, *Doha Work Programme—The Outstanding Implementation Issue on the Relationship between the TRIPS Agreement and the Convention on Biological Diversity—Revision*, Doc IP/C/W/474 and its addenda (also circulated as WT/GC/W/564/Rev.2 and TN/C/W/41/Rev.2), 5 July 2006; WTO–TRIPS Council, *The Relationship between the TRIPS Agreement and the Convention on Biological Diversity: Checklist of Issues*, Doc IP/C/W/420, 2 March 2004; and WTO–TRIPS Agreement, *The Relationship between the TRIPS Agreement and the Convention on Biological Diversity and the Protection of Traditional Knowledge*, Doc IP/C/W/356, 24 June 2002.

[45] See Doc IP/C/W/420 and also more analytically, WTO—TRIPS, *The relationship between the TRIPS agreement and the Convention on Biological Diversity (CBD) and the protection of traditional knowledge—elements of the obligation to disclose evidence of benefit-sharing under the relevant national regime*, Doc IP/C/W/442, 18 March 2005; WTO–TRIPS, *The Relationship between the TRIPS Agreement and the Convention on Biological Diversity (CBD)—Elements of the Obligation to Disclose Evidence of Prior Informed Consent under the Relevant National Regime*, Doc IP/C/W/438, 10 December 2004; and WTO–TRIPS, *Elements of the Obligation to disclose the Source and Country of Origin of the Biological Resources and/or Traditional Knowledge Use in an Invention*, Doc IP/C/W/429/Rev.1, 27 September 2004.

proposal would be achieved through the following options: either by introduction of a new article in the TRIPS agreement; either by addition of an exception to patentability in TRIPS Article 27, allowing Member States to exclude from patentability products or processes that include genetic resources and traditional knowledge obtained in the absence of compliance with international and/or national legislation; or by amending TRIPS Article 29, so that States may require an applicant for a patent to disclose the country and area of origin of any biological resources and traditional knowledge and provide information of compliance with access regulations in the country of origin.

There is finally the moderate position, taken mainly by Switzerland and the EC, favouring enhanced international action in the patent system. More specifically, Switzerland[46] suggests the amendment of the WIPO Patent Cooperation Treaty (PCT) instead of the TRIPS amendment. This would allow Member States to incorporate in their national legislation a requirement for patent applicants to disclose the source of genetic resources and traditional knowledge. In fact, the requirement of disclosure of the source would not be a substantive patent requirement; non-disclosure or false disclosure would only entail procedural legal effects and cause delays in the granting of a patent. To avoid abuse of the disclosure requirement, Switzerland suggests that the invention, where such a requirement applies, be directly based on the genetic resources and/or traditional knowledge accessed by the inventor.

In the same way to Switzerland, the EC and their Member States[47] have adopted a moderate approach, according to which there is no inherent legal conflict between the CBD and TRIPS, since they pursue different objectives. But there is a considerable interaction between the two agreements to be addressed through enhanced international action. In this regard, the patentability requirements of TRIPS Article 27 are quite flexible, allowing

[46] Among a number of communications submitted by Switzerland, see WTO–TRIPS Council, *The relationship between the TRIPS Agreement and the Convention on Biological Diversity (CBD) and the protection of traditional knowledge and folklore and the review of implementation of the TRIPS Agreement under Article 71.1*, Doc IP/C/W/446, 30 May 2005; WTO–TRIPS Council, *Further Observations by Switzerland on its proposals regarding the declaration of the source of genetic resources and traditional knowledge in patent applications*, Doc IP/C/W/433, 22 November 2004; WTO–TRIPS Council, *Additional comments by Switzerland on its proposals submitted to WIPO regarding the declaration of the source of genetic resources and traditional knowledge in patent applications*, Doc IP/C/W/423, 14 June 2004.

[47] WTO–TRIPS Council, *Review of Article 27.3(b) of the TRIPS Agreement, and the Relationship between the TRIPS Agreement and the Convention on Biological Diversity and the Protection of Traditional Knowledge*, Doc IP/C/W/383, 17 October 2002; and WTO–TRIPS Council, *Review of Article 27.3(b) of the TRIPS Agreement, and the Relationship between the TRIPS Agreement and the Convention on Biological Diversity (CBD) and the Protection of Traditional Knowledge and Folklore*, Doc WT/CTE/W/2231, 14 February 2003.

WTO members for considerable leeway to design national patent protection according to their needs, interests and ethical standards, and hence there is no reason to amend TRIPS Article 27.3(b). To tackle the problems that may arise from the implementation of the agreements, the EC recommend action on the means of implementation of the agreements. Capacity building, regulatory/policy development and institutional measures on access to genetic resources and benefit-sharing are vital at the national level. National action can be supported by contractual arrangements between governments, other appointed authorities and private companies. At the international level, the EC recommends policy coherence in all fora that deal with issues related to the interplay between TRIPS, the CBD and the FAO International Treaty.

With regard to the proposals to amend the TRIPS so as to require or enable WTO members to require that patent applicants disclose the source of any genetic material and any traditional knowledge used in the invention, as well as evidence of PIC and fair and equitable benefit-sharing, as a precondition for attainment of the patent, the EC in consonance with Switzerland's view, oppose the inclusion of the source of disclosure requirement in the formal patentability criteria; the legal consequences of non-performance should lie outside patent law, with civil or other administrative action. In the view of the EC and its Member States, a transparent global system, such as enhanced documentation and registers, would allow countries to keep track of all patent applications involving the use of genetic resources and/or traditional knowledge for which they granted access. A simple notification system where patent offices report to a centralized body whenever there has been a PIC and benefit-sharing evidence would be valuable for dissemination of information; the CBD Clearing House Mechanism could serve this role.

### *F. Remarks on the TRIPS Council Agenda*

Whereas the 1947 GATT mandate was limited to 'developing the full use of the resources of the world', the language of the preamble to the WTO Agreement recognizes the complementarity of trade and environmental concerns and declares the '...optimal use of the world's resources in accordance with the objective of sustainable development seeking both to protect and preserve the environment and to enhance the means for doing so in a manner consistent with their respective needs and concerns at different levels of economic development'. Biodiversity concerns are not jeopardized by the WTO rules per se, but biodiversity management heavily depends upon the implementation of these rules.

The Bonn Guidelines encourage countries to take appropriate measures to ensure that genetic material is accessed and used following national PIC and MATs requirements, to guarantee disclosure of information in applications for intellectual property rights, and to prevent the use of genetic resources without PIC. These recommendations to the governments are perceived as incompatible with the current intellectual property system and since the adoption of the Bonn Guidelines, the discussion on the clash between intellectual property regime and the CBD has become more acute.

Obviously, States decided to shift the discussions from the CBD and the WIPO to the WTO, because of the 'teeth' of the WTO system, but in any event the expertise gathered at the WIPO should be given more leverage and effort should be made to avoid duplication of work among the different fora. While the 2004 checklist submitted by the developing countries gave an impetus to the discussions at the WTO, it is ultimately very unlikely that the other members modify their proposals and agree with this stance.

The three basic options to strengthen the synergies between TRIPS and CBD that arise from the discussions are the modification of patent laws; documentation and registration; and the use of different forms of intellectual property, such as geographical indications or *sui generis* forms of protection which include the notion of community based rights. Considering the diverse positions and the improbability of finding common ground in the near future, the most pragmatic, immediate and flexible approach at the moment is to work on the actual system and support information dissemination through databases and registries to combat biopiracy. Still, such an approach is not without impediments: national patent authorities will be called to practice more scrutiny, and most indigenous practices are hard to record, as they are passed on from one generation to the next through oral traditions and are scattered in nature. Despite these shortcomings, building domestic capacity on access and benefit-sharing in provider countries in conjunction with a well-organized information network would facilitate the task of national patent authorities. Registration and documentation processes have already been initiated with the World Bank Indigenous Knowledge Database and a similar initiative by the WIPO to register traditional practices.

## VI. THE WORLD INTELLECTUAL PROPERTY ORGANIZATION

### *A. WIPO Activities on Traditional Knowledge and Genetic Resources*

The WIPO as one of the specialized UN agencies has the mandate to administer, protect and develop intellectual property standards on the basis of cooperation among States and collaboration among international organizations. Its presence at the international discussion relating to genetic resources and traditional knowledge is critical to help clarifying the implications of the discussion on intellectual property. To this end, a number of activities have been initiated, the most important being the establishment of the Intergovernmental Committee on Intellectual Property, Genetic Resources and Folklore (Intergovernmental Committee) and the deliberations within it.

In 1998 the discussions surrounding intellectual property, biodiversity and biotechnology started to identify the difficulties and intensified with more concrete efforts to address the conceptual problems with practical solutions.[48] Fact-finding missions and roundtables with a wide range of stakeholders were organized, regional consultations together with UNESCO, and in cooperation with UNEP, WIPO carried out the On-Site Documentation Project on the Role of Intellectual Property Rights in the Sharing of Benefits Arising from the Use of Traditional Knowledge and Associated Biological Resources, focusing on case studies in India, Mali and Nigeria, where intellectual property rights were used as benefit-sharing tools.[49]

After an exploratory phase, WIPO activities oriented to dissemination of information and education. In this context, education materials were developed for the protection of traditional knowledge under existing intellectual property system and the better management of information and distributed to traditional knowledge holders and intellectual property authorities. Further studies were conducted to assess the differences between modern and customary laws. It was early realized that the collectivity of creation and ownership was one of the main problems for the protection of traditional knowledge.

In September 1999, the WIPO Intergovernmental Committee on Intellectual Property, Genetic Resources and Folklore was established as the specialized body where member states are able to discuss the issues of genetic resources and benefit-sharing; the protection of traditional knowledge, whether associated or not with these resources; and the protection of expressions of folklore. On average, the Committee meets twice a year, and to date, it has convened thirteen sessions.

In response to the Bonn Guidelines, the deliberations within the Intergovernmental Committee focus since 2004 on the Draft Provisions on the Protection of Traditional Knowledge and Traditional Cultural Expressions against misappropriation and misuse.[50] The Provisions, still in

[48] See in sum WIPO–IGC, *Overview of the Committee's Work on Genetic Resources*, Doc WIPO/GRTKF/IC/8/9, 20 April 2005. Also the web site of the Committee at http://www.wipo.int/tk/en (13/10/08).

[49] WIPO/UNEP, *Case Study on the Role of Intellectual Property Rights in the Sharing of Benefits Arising from the Use of Biological Resources and Associated Traditional Knowledge*, jointly produced jointly by the World Intellectual Property Organization (WIPO) and the United Nations Environment Programme (UNEP), prepared by Professor Anil K Gupta, Indian Institute of Management, Ahmedabad, India, 2005.

[50] WIPO *Intellectual Property and Genetic Resources, Traditional Knowledge and Traditional Cultural Resources* (WIPO, Geneva, 2006).

a draft working form, have been designed on the basis of examined community, national and regional practices, and have evolved in consultation with States and indigenous communities, non-governmental organizations and other stakeholders. Traditional knowledge and traditional cultural expressions are dealt with in two separate sets of provisions, to reflect more accurately the different legal and policy particularities that may arise and the different treatment they enjoy in various jurisdictions. However, they are closely linked and the two sets of provisions are complementary to each other.[51]

### *B. Remarks on WIPO's Work*

The creation of the Intergovernmental Committee within WIPO to deal with the technical questions arising from the protection of traditional knowledge and genetic resources has accelerated the rhythm of deliberations at the international level. In parallel to the thorough examination of technical questions of intellectual property nature, the conduct of studies and surveys throughout the world and the establishment of databases on practices, constitute rich sources of information.

The interests represented in this forum are similar to the interests described above for the TRIPS Council and the CBD work. Developing countries have been firm for several years now on the development at the Intergovernmental Committee and adoption of a legally binding international instrument for the protection of traditional knowledge, traditional cultural expressions and genetic resources.[52] But most developed countries argue that deeper examination and better understanding of the issues is needed before moving ahead with such an instrument; or they completely oppose such an international solution, on the basis that the problems can be tackled in a satisfactory at the national level.

Besides, there are still gaps[53] that impede the progress of the deliberations. Among them, a critical issue remains the lack of unanimity and understanding of what constitutes 'traditional knowledge' or 'traditional cultural expression'; the terms have been handled so far in the form of

[51] Draft Provisions on Traditional Cultural Expressions/Folklore and Traditional Knowledge http://www.wipo.int/tk/en/consultations/draft_provisions/draft_provisions.html (13/10/08).

[52] See the Submission of the African Group at the WIPO—IGC Thirteenth Session in October 2008, WIPO—IGC, *African Group Proposal on the Protection of Traditional Knowledge, Traditional Cultural Expressions and Genetic Resources,* Doc WIPO/GRTKF/IC/13/9, 18 September 2008.

[53] See the analyses carried out by the WIPO Secretariat following the mandate of the IGC at its twelfth session, and to be discussed at the thirteenth session in October 2008, WIPO–IGC, *The Protection of Traditional Cultural Expressions—Draft Gap Analysis,* Doc WIPO/GRTKF/IC/13/4(b), 22 August 2008 and WIPO–IGC, *The Protection of Traditional Knowledge—Draft Gap Analysis,* Doc WIPO/GRTKF/IC/13/5(b), 22 August 2008.

working definitions. Also, there are concerns over the ownership of communal resources and over information that has confidential character for many traditional and local communities, and its revelation will cause spiritual or cultural harm. Despite the major efforts to pin down the issues and put forward recommendations, due to the fundamental differences among the international players and unsolved gaps, no significant breakthrough is expected in the coming sessions.

## VII. SUMMARY AND CONCLUSION

Apart from the CBD, three intergovernmental bodies that have to date received most attention with relation to plant genetic resources and form a complicated regime for genetic resources: the FAO with its International Treaty; the WTO and its TRIPS Agreement; and the WIPO.

The FAO was the first international body to set up an international regime for the management of genetic resources; this regime is explicitly dedicated to genetic resources for food and agriculture. In 1983 the Global System for the Conservation and Utilization of Plant Genetic Resources was established and the Commission on Genetic Resources for Food and Agriculture was created. In 1983 the International Undertaking was adopted for the global management of plant genetic resources for food and agriculture, now replaced by the International Treaty. Compared to the IT, the Bonn Guidelines are voluntary and have a much broader scope and apply to all plant genetic resources. The evolution of the Undertaking to the Treaty in the field of food and agriculture is exemplary and it remains to be seen whether the Bonn Guidelines will eventually lead to a binding treaty.

At the same time, the Doha mandate to the WTO–TRIPS Council, to examine the relationship between the CBD and TRIPS Agreement, has triggered intense discussions within this forum on the relationship between the two international agreements, and the related review of TRIPS Article 27 and the protection of traditional knowledge.

Finally, the WIPO has an important role in identifying innovative intellectual property solutions for the technical and empirical work it caries out, especially with relation to the protection of traditional knowledge. As the main administering international forum on intellectual property rights, the WIPO has commenced a range of initiatives for the exploration of possible solutions. It is a very good organization from a technical point of view, but states preferred to transfer harmonization issues of intellectual property to the WTO, a politically strong organization.[54]

[54] Bail in Cottier and Mavroidis (eds) (n 15) 520; See also Doc IP/C/W/370/Rev.1 paras 13–17 where the States' positions on which is the most appropriate forum are summarized.

Political reasons make States negotiate issues in specific fora and there is no answer on which forum would be the most appropriate. This is especially relevant for the WIPO and the WTO; clearly, intellectual property became a core subject at the WTO with the Uruguay Round, but this should not undermine the role of the WIPO. To avoid duplication of efforts and waste of resources, all relevant international bodies acknowledge the need for coordination, but this should be ultimately realized. The conflicting interests and the difficulties in the negotiations indicate the complexity of the issue and imply that states are not yet ready to elaborate a binding treaty on access and benefit-sharing.

# CHAPTER 5

# *Selected Domestic and Regional Implementation*

## I. INTRODUCTION

The Philippines and the Andean Pact have been selected by using the objective criterion of choosing the country and regional cooperation group with the oldest law and policy on access and benefit-sharing, allowing analysis of their regulatory framework over a longer period of time.[1]

The aim of this chapter is to show how, in a given country and region, the access and benefit-sharing type of regulation meets the CBD objectives in manner that corrects the failures of the exclusively contractual approach. Among others, an interesting example is the contract concluded between Bioamazonia, the institution recognized by the Brazilian Government as the manager of the country's natural genetic resources, and the Swiss pharmaceutical company Novartis, where the reaction of the scientific community, legal scholarship, NGOs and indigenous peoples was such that led to the change of the contract from a private-oriented approach to a public law solution. The main arguments raised were related to the violation of Brazil sovereignty, unfair benefit-sharing, unlawful use of traditional knowledge,

[1] S Carrizosa et al (eds) *Accessing Biodiversity and Sharing the Benefits: Lessons from Implementation of the Convention on Biological Diversity* (Gland, Cambridge, 2004); S Pena-Neira, 'Equitably Sharing Benefits Arising from Natural Genetic Resources in Costa Rica and India: Comparative Analysis on the Interpretation at the National Level of Article 15.7 of the Convention on Biological Diversity' (UNU-IAS Working Paper, Tokyo, Japan, 2003); T Kongolo, 'Biodiversity and African Countries' (2002) 24 European Intellectual Property Review 579; P Cullet, 'Property Rights over Biological Resources—India's Proposed Legislative Framework' (2001) 4 Journal of World Intellectual Property 212; G Henne, Genetische Vielfalt als Ressource—Die Regelung ihrer Nutzung (Nomos Verlagsgesellschaft, Baden Baden, 1998) 228 216; N Stoianoff, 'Access to Australia's Biological Resources and Technology Transfer' (1998) 20 European Intellectual Property Review 298; Mugabe et al (eds) *Access to Genetic Resources—Strategies for Sharing Benefits* (ACTS Press, Nairobi, 1997) 94–229; W Lesser, *Sustainable Use of Genetic Resources under the Convention on Biological Diversity: Exploring Access and Benefit-Sharing* (CABI Publishing, New York, 1997) 52 f. See also the following papers in the framework of the First Meeting of Like-Minded Megadiverse Countries, Cancun, Q Roo, Mexico, February 16–18, 2002: J Caillaux and M Ruiz (Peruvian Society of Environmental Law), *Building a Common Agenda First Meeting of Like-Minded Megadiverse Countries and Legislative Experiences on Access to Genetic Resources and Options for Megadiverse Countries*; Ministry of Environment and Forests, Government of India, *Biotechnology and Bioprospecting for Sustainable Development*; and Ministry of the Environment and Natural Resources (Semarnat), Mexico, *Access to Genetic Resources and Fair, and Equitable Sharing of the Benefits.*

and abuse of power, violation of property and of intellectual property rights.[2] The challenge is how to find a balance between the market forces and creativity on the one had, and the protection of the country interests and the rights of indigenous population on the other.

Another issue to be mentioned here the fact that CBD has no retroactive effect, and its provisions on access and benefit-sharing are not applicable to genetic resources accessed before 1993. Apart from the problems related to the quantities of genetic resources accessed before 1993 and currently in gene banks, there is also a gap for the treatment of genetic resources accessed between 1993 and the date a national law operationalized the CBD provisions comes into effect.

## II. NATIONAL ACCESS AND BENEFIT-SHARING POLICY: THE PHILIPPINES[3]

Of the southeast Asian countries that have been experiencing 'biological meltdown', the Philippines is one of the richest in biodiversity.[4] In 2000 French scientists presented themselves as ecotourists and were caught trying to take medicinal plants out of the country secretly.[5] While the CBD came into force in 1993, the Philippines has enacted access and benefit-sharing legislation very quickly and rather comprehensively, to the point that some foreign bioprospectors have sternly criticized Filipino legislation as over-ambitious.[6] Bioprospecting activities have been regulated since long before: established in 1901, the National Museum of the Philippines has been the main agency for supervising collection and taxonomy activities and its mandate was soon broadened to include the arts and science.[7]

In 1987, the Department of Environment and Natural Resources (DENR) was given increased responsibilities with the creation of the Protected Areas Wildlife Bureau. In 1990, government agencies and acade-

[2] S Peña-Nera, C Dieperink, and H Addink, 'Equitable Sharing Benefits from the Utilization of Natural Genetic Resources: The Brazilian Interpretation of the Convention on Biological Diversity' (2002) 6 Electronic Journal of Comparative Law at http://www.ejcl.org/index.html.

[3] Previous version of this section has been published as 'National Measures on Access to Genetic Resources and Benefit-Sharing—the Case of the Philippines (2005) 1 Law, Environment and Development Journal 50. Many thanks go to Mr L Agaloos and Mr AM Taglag from the Protected Areas Wildlife Bureau of the Philippines Department of Environment and Natural Resources for the information and comments they provided.

[4] K Liebig et al, *Governing Biodiversity—Access to Genetic Resources and Approaches to Obtaining Benefits from their Use: the Case of the Philippines*, Reports and Working Papers 5/2002 30 (German Development Institute in collaboration with the South East Asia Regional Initiatives for Community Empowerment (SEARICE), Bonn, 2002); CV Barber and AGM La Viña, 'Regulating Access: The Philippines Experience' in Mugabe et al (eds) (n 1) 116, 121.

[5] R Dalton, 'Bioprospects less than golden' (2004) 429 Nature 598.

[6] Liebig et al (n 4) III.

[7] Barber and La Viña in Mugabe et al (eds) ( n 1) 115, 121.

mic institutions adopted a memorandum of agreement entitled 'Guidelines for the Collection of Biological Specimens in the Philippines', which was mainly an administrative coordination and permit system, with eventually limited impact.[8]

Later, Filipino scientists called for the development of bioprospecting legislation that would put an end to the 'exploitation' of the country's genetic resources by foreign entities and led to the adoption of the Presidential Executive Order 247 (EO 247, 18 May 1995, in force since 18 May 1995) 'Prescribing Guidelines and Establishing a Regulatory Framework for the Prospecting of Biological and Genetic Resources, their By-products and Derivatives, for Scientific and Commercial Purposes'.[9] Further to the mandate of EO 247 Section 15, the Implementing Rules and Regulations on the Prospecting of Biological and Genetic Resources (DENR Administrative Order 96-20, 21 June 1996) were formulated (Implementing Regulations). Together, the EO 247 and the Implementing Regulations established the first detailed legal framework for bioprospecting and access to biological resources.

The Philippines framework soon evolved with the 2001 Wildlife Resources Conservation and Protection Act (Republic Act 9147, 30 July 2001), its 2004 Implementing Rules and Regulations (IRR, Joint DENR-DA-PCSD Administrative Order 01, 18 May 2004), and the Guidelines for Bioprospecting Activities in the Philippines,[10] adopted in 2005.[11] The redrafted access and benefit-sharing regulatory framework amends or repeals any conflicting provisions in the EO 247.

A number of other laws come to complement the access and benefit-sharing framework in the country: the Indigenous Peoples' Rights Act (IPRA, Republic Act 8371, 29 October 1997); the National Integrated Protected Area System Act (Republic Act 7586, 1 June 1992); the Traditional and Alternative Healthcare Act (Republic Act 8423, 9 December 1997); and the National Museum Act (Republic Act 8492, 12 February 1998).

After this background information, the following sections describe the access and benefit-sharing regulatory environment in the Philippines, before and after the revisions incurred with the 2001 Wildlife Act: EO 247, the Indigenous Peoples' Rights Act, and the Wildlife Act (including the Guidelines).

[8] ibid.

[9] K Swiderska, E Dano and O Dubois, *Developing the Philippines' Executive Order No. 247 on Access to Genetic Resources, Participation in Access and Benefit-Sharing Policy, Case Study no 4* (International Institute for Environment and Development (IIED), London, 2001) 7.

[10] The Draft Guidelines were issued according to Section 14 of the Wildlife Act and Rule 14.1 of Section 14 of the Implementing Rules to the Wildlife Act, according to which the Department of Environment and Natural Resources (DENR), the Department of Agriculture (DA) and the Palawan Council for Sustainable Development (PCSD) 'shall issue joint guidelines specific for bioprospecting'

[11] Texts available at the database of the CBD Secretariat at http://www.cbd.int/abs/measures.shtml.

### A. Description

#### 1. *The EO 247 and the implementing regulations*

The preamble of the EO expressly refers to Article 16 of the CBD and to the rights of indigenous cultural communities in the Philippines to preserve any of their knowledge and practices put, directly or indirectly, to commercial use.[12] Certain basic elements in this law involve: the establishment of the Inter-Agency Committee (IAC), a scheme for mandatory research agreements, regulation on achieving PIC from local communities, and requirements on conforming to environmental protection standards.

##### a) Scope of the law

The scope of EO 247 is defined in its Implementing Regulations. It governs all biological and genetic resources in the 'public domain', namely those in the 'water and lands owned by the State that have not been declared alienable and disposable';[13] naturally-growing plants in private lands that are used for bioprospecting with a view to the discovery, research, or use of resources in the pharmaceutical, agricultural, or other commercial sector[14] irrespective of whether the users are foreign or national individuals, entities, or government or private organizations. Traditional uses fall beyond the scope of the law, and thus landraces and animal races are not contemplated by this law.[15]

##### b) PIC procedure

The concept of PIC plays a key role in the application processes for both academic and commercial research agreements.[16] Accordingly, the location and legal status of the collection area determine whose PIC must be solicited. The procedure involves delivery of a copy of the application proposal to the local governmental unit responsible for the collection of resources found in communal land, which must then respond within 60 days; this unit may be represented by the head of the local and/ or indige-

[12] PJ Benavidez, Philippines: Evolving Access and Benefit-Sharing Regulations, in Carrizosa et al (eds) (n 1) 153; CV Barber, L Glowka and AGM La Vina 'Developing and Implementing National Policy Measures for Genetic Resources Access Regulation and Benefit-Sharing' in SA Laird (ed) *Biodiversity and Traditional Knowledge—Equitable Partnerships in Practice*, (Earthscan Publications Ltd., London/Sterling, 2002) 404; Liebig et al (n 4) 31; Swiderska, Dano and Dubois (n 9) 28; Barber and La Viña in Mugabe et al (eds) (n 1) 123; Henne (n 1) 230.

[13] Implementing Regulations Section 2(z).

[14] ibid Section 3.

[15] ibid Section 3.1.b).

[16] EO Section 4 and Implementing Regulations Section 6, and Implementing Regulations Section 2.1.w (PIC definition) and Sections 6.1.3 and 7.

nous cultural community or communities that may be affected, the protected area's management board, or a private landowner.

Prospecting of genetic resources within the borders of ancestral lands and indigenous cultural communities shall be allowed only once the PIC of the concerned communities has been obtained in accordance with their customary laws. Indigenous cultural communities or indigenous peoples are defined as a:

> homogenous society identified by self-ascription and ascription by others who have continuously lived as community on mutually bounded and defined territory, sharing common bonds of language, customs, traditions and other distinctive cultural traits, and who, through resistance to the political, social and cultural inroads of colonization, became historically differentiated from the majority of Filipinos.[17]

c) Academic Research Agreements (ARAs) and Commercial Research Agreements (CRAs)

Once the PIC has been obtained and the application for carrying out bioprospecting activities has been accepted, the nature of the activities envisaged defines whether the contracting parties sign an Academic Research Agreement (ARA) or a Commercial Research Agreement (CRA) and the EO 247 stipulates their terms.[18] ARAs deal with prospecting biological and genetic resources for academic purposes; and actually only 'duly recognized Filipino universities and academic institutions, domestic governmental and intergovernmental entities' can apply for ARAs. The EO assumes that all bioprospecting agreements other than those concluded with domestic research institutions and domestic governmental/intergovernmental entities have commercial ends. In fact, this operates as a precaution against ARAs that act as a cover for any future commercial use made of the results of academic research, for activities that are carried out abroad.

CRAs, on the other hand, deal with research and collection activities intended, whether directly or indirectly, for commercial purposes and thus encompass all agreements stipulated with private persons, corporations, and foreign or international entities. In cases where a domestic academic bioprospector detects the possibility of marketing the results of research, s/he must apply for a CRA.[19]

[17] ibid Section 2.1.r.
[18] EO Section 3 and 5 and Implementing Regulations Section 8.
[19] ibid Section 5(n).

d) The Inter-Agency Committee (IAC)

The EO adopted an inter-agency approach by establishing the Inter-Agency Committee,[20] with the mandate to coordinate the processing of bioprospecting applications and access and benefit-sharing procedures in the country, and to discuss any relating institutional, political, and technological developments. Its members represented the Department of Environment and the Ministry of Foreign Affairs, the sectors of science and technology, agriculture, and health, and the National Museum; there was also a person from an NGO active in biodiversity protection and a person from the 'Peoples' organization' of indigenous cultural communities. Despite the good intentions to have multiple stakeholders in this body, lack of clarity with regards to the responsibilities of the IAC members and administrative difficulties led soon to the abolition of the agency.

e) Environmental protection requirements

All bioprospecting activities and their results must not, whether directly or indirectly, harm biodiversity, ecological balances, or the inhabitants of the area where resources are being collected. To this end, an Environmental Impact Assessment (EIA) is required to ensure conformity with these requirements is mandatory.[21]

f) Other aspects

Any decision may be appealed within 30 days.[22] Actually, prospecting activities without an agreement are considered criminal acts and any behavior that contradicts the agreement is a reason for cancellation/revocation in favour of the government, confiscation of collected material, forfeit of bond, and perpetual ban from prospecting activities in the Philippines without prejudice to administrative sanctions.[23]

### 2. *The 1997 Indigenous Peoples' Rights Act*

The Indigenous Peoples' Rights Act (IPRA) was adopted in 1997 as a way of recognizing, protecting, and promoting the rights of Indigenous Cultural Communities and Indigenous Peoples. It is one of the most comprehensive attempts at introducing the principles in Article 8j of the CBD into national legislation.

[20] EO Section 6 and Implementing Regulations Section 10, and Implementing Regulations Section 11.
[21] ibid Section 6.1.4. [22] ibid Section 13. [23] ibid Section 14.

The Act created the National Commission on Indigenous Peoples (NCIP).[24] It upholds indigenous ownership rights over land and resources, including the right to ownership and development of lands and resources, and the right to regulate the entry of migrant settlers and organizations.[25]'Ancestral domain'[26] is defined by conferring property rights to indigenous communities over lands, inland waters, coastal areas, and natural resources contained therein. In this context, indigenous customary laws and existing property rights regimes are expressly recognized.[27] The Free and Prior Informed Consent (FPIC) of members of the indigenous communities is required for any activity conducted by outsiders on an ancestral domain.[28]

The IPRA recognizes full ownership and control of the indigenous peoples over their cultural and intellectual rights, over their sciences and technologies and derivatives. FPIC of the indigenous peoples is necessary for third parties to obtain any sort of intellectual property relating to specific elements of indigenous culture, knowledge and activity. In the event of violation of this provision in a foreign country, the intellectual property right obtained will be considered illegal or not honored in the Philippines.[29]

At the same time, the recognized rights over tangible and intangible property as listed above are conditional on the obligation to maintain ecological balance and restore denuded areas.[30] The 'prior informed consent' mentioned in the EO and the 'free prior informed consent' included in the IPRA were not identical, but the IPRA provisions were the modus operandi for all bioprospecting activities in ancestral domains. This overlap has now been explicitly resolved with the 2005 Guidelines, presented in the following section.

### *3. The Wildlife Act and its Implementing Rules (Bioprospecting Undertaking)*

The Wildlife Resources Conservation and Protection Act entered into force in 2001 with the primary objective of conserving and ensuring the sustainability of all wildlife resources and habitats in the Philippines.[31] The Act defines bioprospecting as research, collection, and utilization of biological

[24] IPRA Chapter VII.

[25] ibid Section 7.

[26] See also the definition of 'ancestral domains' in the EO Implementing Regulations Section 2.1.c).

[27] IPRA Section 2.

[28] ibid Section 35, where it is stated that 'access to biological and genetic resources and to indigenous knowledge related to the conservation, utilization and enhancement or these resources, shall be allowed within ancestral domains of the indigenous cultural communities/ indigenous peoples only with prior informed consent of such communities, obtained in accordance with customary laws of the concerned community'

[29] ibid Section 34.

[30] ibid Section 9.

[31] Wildlife Act Section 2.

and genetic resources solely for commercial purposes[32] and deems such activities as legal once the interested party has formally declared in a 'Biodiversity Undertaking' its willingness to abide by the terms and conditions set by the Secretary (of the DENR or the Department of agriculture or both)[33] for the protection of biodiversity. The Biodiversity Undertaking requires the interested party to obtain the PIC of concerned indigenous communities and, if the applicant is a foreign individual/entity, a local institution must become involved in the agreement.

The Guidelines came to address the shortcomings of the EO 247 and attempt to streamline the access and benefit-sharing procedure in the Philippines, facilitate compliance, and establish a cost-effective, transparent, and standardized system. Their preparation was the result of a multi-stakeholder process, where three public consultations involving government agencies, academic institutions, NGOs, indigenous peoples' organizations, and representatives of other private and public sector entities took place.

The Guidelines infer the following main amendments to the EO 247: bioprospecting agreements are now of an exclusive commercial nature, and hence procedures for academic research are simplified; the Inter-Agency Committee is abolished and the role of the Secretary (defined depending on the nature of the resources) is strengthened; and detailed guidance on benefit-sharing is implemented. From a procedural point of view, the review and evaluation time for bioprospecting applications is significantly shortened; the fees for local commercial researchers with the aim of promoting research and development in the country are reduced; and a local individual/entity must always be involved in bioprospecting activities. A uniform format is finally laid out of Material Transfer Agreements, for requesting and submitting PIC, etc.[34]

### a) Scope

The Guidelines apply to all biological and genetic resources found in the country: wildlife, private lands, *ex situ* collections, protected areas, and ancestral domains.[35] They establish a uniform procedure for evaluating and granting access to biological and genetic resources, independent of the government agency managing the resources. The Guidelines apply only to Commercial Research Agreements, as defined in the EO 247. For Academic

[32] ibid Section 5(a).

[33] Wildlife Act Section 4 confers jurisdiction to the DENR over all terrestrial wildlife and to the Department of Agriculture over aquatic wildlife, except for the Province of Palawan, where because of the existing special regime, jurisdiction is vested to the Palawan Council for Sustainable Development. The distribution of jurisdictional powers is elucidated in the Implementing Rules.

[34] See annexes of the Guidelines.

[35] Guidelines, Section 2.

Research Agreements, the applicant must execute an Affidavit of Undertaking or Memorandum of Agreement and the concerned authorities issue the gratuitous permit.[36] The Guidelines do not apply in case of collection of resources for application in traditional practices.[37]

b) PIC procedure

Compared to EO 247, the changes inferred relate to the stages of notification and public consultation in the process of the PIC certificate:

i) The proposed user of the resources has to inform interested parties of the intention to carry out bioprospecting activities through a Letter of Intent, rather than through the media;
ii) The requirement has been added that the summary of the research proposal must now be provided to the affected community in a language/dialect that is understandable to them;[38]
iii) The Guidelines expressly refer to the FPIC process of the IPRA in cases where consent needs to be obtained from indigenous peoples;[39]
iv) Concerned agencies or private owners have 30 days to issue the PIC certificate, half the time stipulated in the EO to the community.

c) Detailed benefit-sharing provisions

The Implementing Regulations of the EO 247 provided definitions of benefit-sharing and equitable sharing, and an indicative list of results, such as royalties and capacity building,[40] but did not actually regulate benefit-sharing. The Guidelines contain detailed provisions on the benefit-sharing terms of bioprospecting agreement.[41] Fully in the spirit of the CBD, equitable sharing means that monetary benefits reach the concerned local community '*and are used solely for biodiversity conservation or environmental protection, including alternative or supplemental livelihood opportunities for community members*'.[42] Also, the Guidelines put in place a mechanism for monitoring with a checklist of performance indicators for benefit-sharing.[43]

### *B. Analysis and Evaluation*

The Philippines access and benefit-sharing legislation is evaluated here by observing the scope, the representation of stakeholders' interests, the application of PIC procedures, and benefit-sharing stipulations.

[36] Wildlife Act Section 15 and Implementing Regulations.
[37] Implementing Regulations Sections 4 and 5.
[38] Guidelines Chapter 5.
[39] ibid Section 4 and 12.
[40] Implementing Regulations 2.1. e) and n).
[41] ibid Chapter VI.
[42] Guidelines Sections 19.1 and 19.2.
[43] ibid Section 23 and Annex V.

*1. Scope—*Ex situ *collections*

In absence of clarity in the EO and its Implementing Regulation, the Guidelines now extend the jurisdiction of access and benefit-sharing provisions to include *ex situ* collections, such as those held by the International Rice Research Institute[44] based in the Philippines, when these resources are used for commercial purposes. In this respect the new framework goes further than the CBD or the Bonn Guidelines. In fact, the role of the IRRI is special in that it is a trustee rather than beneficial owner of the plant genetic material within its confines.[45] Obtaining access to IRRI collections, now requires the conclusion of a standard Material Transfer Agreement that includes a clause prohibiting recipients of IRRI germplasm from claiming either ownership of the material or intellectual property rights over that germplasm or related information.[46]

*2. Stakeholders' representation*[47]

Stakeholders that may be involved in bioprospecting agreements include the state, local/indigenous communities, scientific institutions, representatives of the industrial sector, and NGOs.

a) The State

The State retains a strong role in monitoring and enforcing the law. After the decision to disband the Inter-Agency Committee, the State is now represented by the responsible agencies, depending on whether the resources to be accessed belong to terrestrial or marine biodiversity.

b) Local/indigenous communities

Local and indigenous communities become stakeholders whenever bioprospecting activities take place on lands that are not designated protected areas or are not privately owned. Prior informed consent procedures and benefit-sharing provisions consider their participation in the process. In the Philippines, there are two main types of local (non-indige-

[44] See the web site at http://www.irri.org/ (11/11/08).

[45] C Fowler, 'The Status of Public and Proprietary Germplasm and Information: An Assessment of Recent Developments at FAO' IP Strategy Today No 7-2003, 1.

[46] H Leung, GP Hettel and RP Cantrell, 'International Rice Institute: Roles and Challenges As We Enter the genomics Era' 7 TRENDS in Plant Science 139 (2002) and K Kate and A Collis, *Benefit-Sharing Case Study—The Genetic Resources Recognition Fund of the University of California, Davis*, 5 Submission to the Executive Secretary of the Convention on Biological Diversity by the Royal Botanic Gardens, Kew, 1998.

[47] Liebig et al (n 1) 34 f and 63; Swiderska, Dano and Dubois (n 9) 7 and 12.

nous) political organization: *barangays* (communities of approximately 500 households) that are represented by a *captain*, and municipalities, represented by a mayor. The latter are assigned the responsibility of deciding whether to sign a PIC certificate or not, on the basis of case-by-case analyses. The active participation of the communities' members rests largely on the local political elite's commitment to involving his/her constituency and on awareness-raising activities carried out by NGOs.

When bioprospecting activities take place on ancestral lands, the situation is different. The Philippines hosts approximately 4.5 million indigenous peoples representing more than 70 ethno-linguistic groups. Indigenous peoples are distinguished for their particular cultural lifestyles and their attitude to using and managing natural resources that privileges sustainability and that is based on a holistic perception of the land. Due to rather negative experiences with outside interventions, such as logging and dam-building, which occasionally led to violations of rights and exploitation of the affected communities, indigenous communities appear generally skeptical of the potential benefits of bioprospecting.

c) Scientific community

Filipino scientists have long recognized that the exploitation of biodiversity has rarely been beneficial for the country. It was a group of scientists that actually triggered off the development of EO 247, by mobilizing political and legal actors against the exploitation of the country's natural resources by foreigners for the sake of domestic technological development. However, since the EO 247 has entered into force, only three research agreements have been signed (two CRAs and one ARA) and all with the University of the Philippines in Diliman, Quezon City.

d) Industry

Domestic industrial actors in the Philippines have not been involved in bioprospecting, because access and benefit-sharing regulations are not very relevant for their activities.

e) NGO community

The NGO community of the Philippines is very dynamic and its role has been instrumental in empowering local communities and promoting sustainable development and biodiversity conservation in the country. On access and benefit-sharing, NGOs tend to follow one of two extreme positions: the first rejects any bioprospecting activities in the country, claiming that they constitute biopiracy and capitalist exploitation; the second is generally supportive of bioprospecting activities, as long as they abide by existing rules. Supportive NGOs participate actively in the implementation and enforcement of access and benefit-sharing legislation and used to be very involved in the Inter-Agency Committee.

### *3. Application procedures for agreements and obtaining PIC*

The issue of PIC is absolutely central to Filipino legislation on access and benefit-sharing. The process is based on public notification and sector consultation with the concerned local/indigenous cultural communities to secure consent and thus meets the requirements of the Articles 8j and 15.3 of the CBD. The principles of local community participation and protection are reflected in both the EO 247 and with slight revisions in the Guidelines.[48]

Actual implementation of existing PIC mechanisms remains a difficult mission; if the communities are not fully aware of their rights and/or of the enforcement procedures, their effective participation cannot be ensured, where 'effective' is understood as informing and instructing the affected populations in a language they understand, as well as allowing them to reflect and make a decision. Finally, consultative procedures lack democratic endorsement, if certain tribal leaders and families enjoy exceptional privileges and can significantly influence the community decisions.

Due to the problems encountered in obtaining PIC from indigenous communities, it is almost presumed that it is impossible to carry out any bioprospecting activity on ancestral domains/lands, and when bioprospecting is about to take place in these lands, strong evidence must be submitted to the authorities demonstrating that the affected indigenous community/ies actually want the activity to take place.[49] Obtaining PIC is rather complex and discourages users, who are asked to cooperate on-site with governmental entities, NGOs, and peoples' organizations so as to identify the communities' representatives and, when necessary, to apply the relevant customary law.

### *4. Sharing the benefits*

The Guidelines build upon the framework of the EO 247 but provide for terms on how any potential monetary or non-monetary benefits reach the source communities. Also, benefits should be used to benefit conservation efforts. The two agreements stipulated to date—between the University of the Philippines and the University of Utah, and between the Pascual Laboratory and a British pharmaceutical company—show that there have

[48] Liebig et al (n 1) 42.
[49] Barber and La Viña in Mugabe et al (eds) (n 1) 129.

not been long-term benefits yet and that the short-term benefits go first to the scientific community of the Philippines.[50] In the case of the collaboration between the Pascual Laboratory and the British pharmaceutical company, a joint venture was created.[51]

### C. *Final Remarks*

In theory, the Philippines framework on the application, PIC, and benefit-sharing procedures are designed to guarantee transparency, stakeholder participation, and equitable sharing of the benefits. In practice, though, these laws have been criticized for their long and complex procedures and there has been a very low record of applications for research agreements.[52] Also, insufficient funding and difficulties in defining representation among indigenous peoples at the time of development of the EO 247 hindered participation of all stakeholders.

From the viewpoint of industry, the main bone of contention is the procedure for obtaining PIC, which is particularly difficult because the Philippines has placed solid protective measures (such as the IPRA) as a means of safeguarding its indigenous communities. A step that the Philippines could take is to promote a unified, regional system in the ASEAN context. The need for a regional regulatory instrument was emphasized in the Melaka Accord (Malaysia), the outcome of the Eighth Asian Symposium on Medicinal Plants, Spices and other Natural Products (ASOMPS VIII), 12–16 June 1994. Although access and benefit-sharing legislation is underway in a number of ASEAN countries, the Philippines regulatory framework sets higher performance standards for industry and research institutions. This means that the Philippines are at a competitive disadvantage *vis-à-vis* neighboring countries because users will prefer to invest in less 'complicated' countries with biologically similar environments. Such a regional effort is studied in the next section.

## III. REGIONAL ACCESS AND BENEFIT-SHARING POLICY: THE ANDEAN PACT

At present, the tendency for regional cooperation areas is apparent worldwide. These areas start out with economic purposes, but in many cases they extend their competencies in various areas, the most characteristic example being that of the EU. According to the CBD, States are sovereign over their natural resources and are authorized to determine access to genetic

[50] Liebig et al (n 1) 47.

[51] ibid.

[52] Liebig et al (n 1) 23 and 28; Swiderska, Dano and Dubois (n 9) 7.

resources. States are also members of free-trade areas or customs unions. The economic character of the access and benefit-sharing regulation suggests that the legislative and implementing competencies may be entrusted to the regional trade integration unions to which states belong.

In 1969, Bolivia, Chile, Colombia, Ecuador and Peru formed the Acuerdo di Cartagena (Andean Pact).[53] Chile withdrew its membership in 1976. Also, Venezuela was a member from 1973 until 2006. The development of a common regime for the regulation of access to genetic resources and benefit-sharing in the Andean Community was inaugurated in 1993 with Decision 345 on Plant Breeders' Rights,[54] a hybrid of UPOV 1978 and UPOV 1991. Since February 1996, the Andean Community is the first regional integration and trading bloc to have established a common access regime, which is the second most comprehensive access and benefit-sharing regulation after the Philippines. The negotiations started in 1994, and two years after, the members of the Junta del Acuerdo de Cartagena (JUNAC), the technical body of the Andean Pact, adopted unanimously the draft Decision 'Common Regime on Genetic Resources' (Proyecto de Decisión 'Régimen Commún sobre los Recursos Genéticos'), which became legally binding with its publication in the Andean Pact Gazette on 17 July 1996.

A regional approach offers many advantages to developing countries rich in biodiversity with environmental and biological similarities. Decision 391 is an exemplary regional effort, but suffers from a number of drawbacks. In the following sections, light will be shed on Decision 391, and a few words shall be drawn with respect to Decision 486 on the common intellectual property rights regime.

## *A. Description*

### *1. Decision 391: Common Regime on Access to Genetic Resources*[55]

At the preparation stage,[56] government experts determined the outline and

[53] The official web site of the Comunidad Andina is http://www.comunidadandina.org/ (7/12/08).

[54] See Decision 345, Transitional Provision Three, where the Member States engage in approving, before 1994, common provisions on access to biogenetic resources and biosecurity pursuant to the CBD provisions.

[55] This part draws on Chaves, the Andean Pact and Traditional Environmental Knowledge' in NP Stoianoff (ed) *Accessing Biological Resources—Complying with the Convention on Biological Diversity* (Kluwer Law International, New York, 2004); D Mongollón-Rojas, 'The Preservation of Local Biodiversity Inheritance and Indigenous People's Knowledge Proprietorship in the Venezuelan and Andean Community Legislation' (2002) 5 The Journal of World Intellectual Property 539; M Ruiz, 'Research in Protected areas in the context of Decision 391 of the Andean Pact and Decision 391: The Common Regime on Access to Genetic Resources in the Andean Pact' in Laird (n 12); Henne (n 1) 236 f and M Ruiz Muller,

the basic elements of the decision in the framework of five Governmental Expert Meetings, while three workshops were organized to ensure participation by other interested parties. To avoid conflicts among States, the Andean Pact Council entrusted the preparation of the first technical paper to NGOs with a good reputation for their technical and legal expertise in environmental matters.[57] Due to ideological divergences and disagreement among assigned NGOs, Colombia and Venezuela, still a member at the time, took the initiative to submit a joint country proposal leaving out most of the original concerns, such as the rights of traditional communities to veto any access activity, when their collective rights are being violated by that activity. As a result, NGOs missed a very important opportunity to ensure that the experts' work was based on a non-governmental proposal, and since then their participation in the regulatory processes has been limited.

Some specific elements of Decision 391 are analysed in the forthcoming paragraphs. The Decision is divided into several sections: the first one establishes definitions, the second and third define the Decision's purpose, aims and scope, the fourth section sets the interpretative principles of the legal framework, such as state sovereignty over genetic resources and by-products, recognition of traditional knowledge, regional cooperation, precautionary principle, and the fifth section provides for the access process.

a) Scope[58]

Decision 391 applies to *ex situ* and *in situ* conserved genetic resources originating from Andean Community countries, to their by-products and the associated intangible components; such resources may have been physically obtained and used for purposes of research, biological prospecting, conservation, industrial application and commercial use. This list of enumerated activities is not exclusive and thus all activities directly related to the use of genetic resources, their by-products and intangible components are encompassed.

A by-product ('producto derivado'), for the Decision, is a molecule, a combination or mixture of natural molecules, including crude extracts of live or dead organisms of biological origin that come from the metabolism of living beings. Further, intangible components ('componente intangible') are all know-how, innovation or individual or collective practice, with a

'The Legal Framework on Access to Genetic Resources in the Americas' *Programa de Asuntos Internacionales, Diversidad Biologica,* Sociedad Peruana de Derecho Ambiental, Lima, November 1998; Ruiz Muller in Mugabe et al (eds) (n 1).

[56] For the history of the Decision, see Chaves in Stoianoff (ed) (n 55) 237 f and Ruiz in Mugabe (ed) (n 1) 188.

[57] IUCN-World Conservation Union and the Peruvian Society for Environmental Law (Sociedad Peruana de Derecho Ambiental-SPDA).

[58] Decision 391 Articles 1, 3 and 4.

real or potential value, that is associated with the genetic resources, its by-products or the biological resource that contain them, whether or not they are protected by intellectual property regimes.[59] The supplier may be a farmer, indigenous person, investigator, or anyone who contributed to the development of the primary genetic resource. The definitions of intangible component and its supplier is broader than CBD 8j. The Decision does not specify measures for the protection of intangible components, and only sets the basic guidelines to be further elaborated through national legislation.[60] Access to intangible components requires the conclusion of an ancillary contract annexed to the access contract, which would detail the terms for fair and equitable use of the component.[61]

During the negotiations, an extension of the decision had been proposed to include synthesized products, namely substances obtained through artificial processing or of information from other biological molecules. This line of argument eventually failed. From a practical point of view, there are serious difficulties in controlling and monitoring synthesized products; and in fact, it is questionable whether source countries may retain sovereign claims over synthesized products and regulate access to them.[62] The final draft text of the Decision excludes synthesized products; synthesized products may be subject to contracts between their providers and users, where the terms for their use are set, but monitoring is still an issue.

Human genetic resources are excluded from the scope of the Decision. This exclusion does not however make access to human genetic resources impossible.[63] The exchange of genetic resources, by-products and intangible components among indigenous and local communities for their own consumption according to customary practices falls out of the scope of the

[59] Decision 391 Article 1; the Eighth Temporary Provision requires that Member States present national studies and establish a special regime or a harmonization regulation for the protection of the intangible components in accordance with ILO 169 Article 7 and the CBD.

[60] Decision 391 Article 7. National legislation regarding indigenous peoples varies considerably from country to country in the Pact. Colombia has made significant advances at the constitutional and legislative level in the recognition of specific rights and Peru is discussing a national register for the knowledge, innovations and practices of the local and indigenous communities assisting the State in fair and equitable sharing of the benefits, coupled with a licensing regime based on a *sui generis* know-how license, see Consorcio GTZ/FUNDECO/IE (2001) and M Ruiz Muller, 'Alcances iniciales sobre un registro de conocimientos, innovaciones y practicas de comunidades indigenas asociados a la biodiversidad' (2002) *Serie de Política y Derecho Ambiental de la SPDA Sociedad Peruana di Derecho Ambiental.*

[61] Decision 391 Articles 34–35.

[62] Third Governmental Experts Meeting in May 1995 in Cochabamba-Bolivia, see Ruiz in Mugabe (n 1) 191.

[63] See also Decision 344 of the Andean Pact on a Common Regime on Industrial Property, Article 7(d) establishes that 'inventions related to materials which form the human body or genetic components of it are not patentable', but this provision does not prohibit access to the human genetic material nor does it prevent patenting beyond the Andean countries' jurisdictions.

Decision. The Decision distinguishes between indigenous, Afro-American and local communities, yet it defines them together as groups with social, cultural and economic conditions which differentiates them from other sectors of the national community that is governed totally or partially by its own customs or traditions or by special legislation and that irrespective of its legal status conserves its own social economic, cultural and political institutions or a part of them.

In some cases, the access to genetic resources and their by-products can be limited through the national rules in case of endemic, rare or endangered with extinction species, subspecies, varieties or races; of fragile ecosystems that could worsen as a result of access activities; of adverse effects of access activities on human health or elements essential to the cultural identity of nations; of undesirable or not easily controlled environmental affects or access activities on the ecosystems; the danger of genetic erosion caused by access activities; of regulations of bio-security; or genetic resources or geographic areas rated as strategic.[64] The list of the limitations to access is exclusive and comprehensive. But, it does not define when a resource or a geographic area should be defined as 'strategic'.

This provision on all the limitations to access clarifies the situation in relation to the CBD: the use of genetic resources is subordinate to the conservation of the genetic resources. In any case, the decision is left to the discretion of the Member Countries whether these enumerated cases will be limited from access ('Los Países Miembros *podrán...*'—'Member Countries *may* establish...').

On the whole, because of the inclusion of the by-products and the intangible component, and also the non-exclusive character of the use activities, the scope of Decision 391 is comprehensive and much more extensive than the scope of the CBD. Where there is no access to genetic resources falling within the scope of the Decision 391, the rules for free exchange of goods and services and the CITES rules over health, food safety and biotechnology shall apply.[65]

b) Legal status of genetic resources[66]

Member Countries have sovereign rights over the genetic resources and their by-products, which are affirmed as national goods *or* national property. Genetic resources and their by-products are inalienable and subject neither to prescription nor to seizure or similar measures. At the same time, this Andean Community regulation does not affect the national property

[64] Decision 391 Article 45.
[65] ibid Article 14.
[66] ibid Articles 5 and 6.

regime for the biological resources, for the land where biological resources are found, and for the associated intangible characteristics.

Decision 391 makes a clear distinction between genetic and biological resources. Biological/natural resources can be negotiated by anyone having a right over them; but only the State has rights over genetic resources and the State alone can negotiate benefit-sharing provisions from the use of genetic resources. Accordingly, any right given to a person or entity, such as a forest concession, over natural resources, does not imply rights over genetic material found in the natural resources or the area over which the concession extends.

c) Institutional framework[67]

The 'Autoridad Nacional Competente' of every member country is the National Focal Point of the Andean countries for access and benefit-sharing, with the mandate to oversee all activities related to access to genetic resources and by-products, at all stages from the application to the supervision of the contract; its performance or termination; and to ensure the rights of suppliers of biological resources.

Further, it issues the necessary administrative provisions. It keeps a directory of technical experts, it publishes all actions from access application until the contract, and includes all national inventory of genetic resources and their by-products. It supervises the state of conservation of the biological resources containing genetic resources, it coordinates with all liaison institutions, and keeps in continuous contact with the competent national offices for industrial property.

At the regional level, the Decision creates an advisory body, the 'Comité Andino sobre Recursos Genéticos'. It is composed of the directors, representatives, or advisors of the Competent National Authorities, and also of the representatives of other interested sectors. It issues recommendations for the best possible fulfillment of the common regime on access and for related technical matters submitted by the Member Countries; for the establishment of an Andean information network on applications for access contracts, for the strengthening of the capacity in research and technology transfer, and for the standardization of documents in the classification of resources and the legality of their access. Finally, it coordinates administration, surveillance, control, and cooperation with regard to access to the resources that exist in two or more member countries.

[67] Decision 391 Articles 1, 50 and 51.

d) Procedural matters[68]

The access procedure comprises the application, the publication of the application, the evaluation of the application, the permission to proceed, the completion of the access contract with various accessory contracts, and the issuance and publication of a related resolution and of the declaratory registration of all the access-related acts.

The potential user files an application, accompanied by the project proposal, for access to the 'Autoridad'. Data must be provided on the applicant and on the supplier of the biological and genetic resources, their by-products, or of the associated intangible component, including their present and potential uses that the applicant is in a position to know at the moment of the application. Also, data must be provided on the person or national institution that scientifically or culturally supports the applicant, on the responsible persons for the project, its working group, and the activity and the location or the area where the access is to be carried out.[69] Under circumstances, these data may be given confidential treatment.[70]

The national law of the Member Countries or the Autoridad, if it deems it necessary, can require that the applicant shall comply with environmental provisions in effect. The related procedures are independent from those stipulated in the Decision and may be started beforehand. Nevertheless, they must be concluded before the time of the inspection of the site of prospecting and its evaluation by the Autoridad.[71]

The access contract is concluded between the State, represented by the Autoridad and the applicant.[72] The access contract, the application for access, and the accessory contracts must contain provisions related to the participation of the nationals in the region where bioprospecting will take place; the support for research that contributes to conservation and sustainable use of the biological diversity in the country or region of origin; the technology and know-how transfer that is culturally, socially and environmentally healthy and safe; the strengthening of national or regional institutions with respect to genetic resources and their by-products; the strengthening of the communities in relation to the intangible components

[68] ibid Articles 16–44. [69] ibid Articles 22 and 26.

[70] ibid Article 19. The data that the applicant provides may be given confidential treatment after his request, unless the public knowledge of this data and information is necessary to protect the social interest or the environment. For this reason, at his/her request, the applicant should state clearly the grounds of his petition for confidentiality, accompanied by a non/confidential summary that will be part of the public file. The information or documents, however, related to the applicant, the supplier, the supporting institution, and the access method cannot be treated confidentially.

[71] ibid Article 31.

[72] ibid Article 32.

of the resources; and also, information about the status of the science and other knowledge that contribute to a the better understanding of genetic resources, their by-products and synthesized products or intangible components.[73]

Part of the access contracts are the ancillary contracts between the applicant and the owner, possessor, or manager of the land, where the genetic resource is allocated and the private or public support institution, which participates.[74] When access is requested to resources with an intangible component, the access contract shall incorporate as annex the fair and equitable sharing of the profits from use of those characteristics.[75] To be considered executed, the contracts must be published.[76] Contacts that contradict the access common regime are null and void. The Decision provides for the type and measure of the sanctions the Autoridad will apply in case of access without contract and therefore unauthorized access, or in case of breach of the contracts. The proposed sanctions include fines, confiscation and closing-down of establishments.[77]

### 2. *Decision 486: Common Intellectual Property Regime*[78]

According to Decision 391, Member Countries do not recognize any rights, including intellectual property rights, over genetic resources, by-products or synthesized products and associated intangible components obtained through an access activity in violation of the common regime on access.[79]

Decision 486 of 2000 aims at establishing a common regime on intellectual property for the Andean Community Member Countries: Member Countries accord intellectual property rights and at the same time safeguard and respect their biological and genetic heritage and the knowledge of the traditional/local communities. In practical terms, Decision 486 requires that the granting of every patent application based on such heritage or knowledge, is subordinate to the international, national and Andean law (explicitly referring to Decision 391) with regards to the acquisition of the material. It is the first time that an intellectual property law makes reference to the term 'biological patrimony', which reinforces the idea of sovereignty over these resources. The Decision further recognizes the role of traditional knowledge and practices in biodiversity conservation and the authority and right of indigenous peoples over their collective knowledge.[80]

[73] ibid Article 17.
[74] ibid Articles 41–44.
[75] ibid Article 32.
[76] ibid Article 38.
[77] ibid Articles 46 and 47.
[78] Consorcio GTZ/FUNDECO/IE (n 60) 65 f.
[79] Decision 391 Second Complementary Provision.
[80] Decision 486 Article 3.

Following these principles, Decision 486 requires, as part of the patent application, a copy of the contract of access and the license or authorization of the use of the traditional knowledge with a view to ensuring and controlling the legality of the access where necessary. Failure to submit these documents results in the consideration of the application as 'abandoned';[81] if a patent has already been granted and it is later proved that the genetic resources or the traditional knowledge from one of the Andean countries was acquired in a illegal manner and without complying with the respective procedures, the patent is declared null and void.

### *B. Analysis and Evaluation*

Member Countries of the Andean Community had no national ABS policies before Decision 391 was adopted. The Decisions of the Community have the form of 'Common Regimes', which set a minimum standards framework for its Member Countries. They do not require particular action from the Member State to integrate them into their legislature and they are not subject to any national constitutional control;[82] once approved by the minister, decisions of the Andean Community are directly applicable and binding upon the Member States.[83] However, the Decisions may need to be further substantiated at the domestic level through administrative regulations and structures.

With regards to Decision 391, the intention was that Member Countries would only need to develop few additional dispositions and regulations, but in effect, technical, social, political and institutional concerns and limitations made the development of parallel national benefit-sharing legislation in most Andean countries a compelling task.[84] For example, the rules of the Decision defining the time limits of the evaluation of the application for access (eg 'maximum 40 days') and those on compliance monitoring and control are imprecise. Now Decision 391 stands between the CBD framework rules for ABS and their national transposition.

[81] ibid Article 39.

[82] The Acuerdo di Cartagena, an international treaty, presumes that any Andean Decision would not violate any Member State constitution; the Acuerdo has been ratified by the national congress of each Member State and constitutes part of their national legislation, and hence Andean Decisions are part of this treaty, see J Chaves, 'The Andean Pact and Traditional Environmental Knowledge' in NP Stoianoff (ed) *Accessing Biological Resources—Complying with the Convention on Biological Diversity* (Kluwer Law International, New York, 2004) 240.

[83] ibid and M Ruiz Muller, 'Access regime for Andean Pact countries: Issues and experiences' in Mugabe et al (n 1) 198, fn 2.

[84] S Carrizosa, 'Diversity of Policies in Place and in Progress' in Carrizosa et al (eds) (n 1) 9 f.

Colombia designated a competent authority in 1998, but its attempts to negotiate an access application by direct application of Decision 391, failed;[85] it is currently working on a proposal for a national ABS policy.[86] In Bolivia, since 1997 there is a national law for the Decision 391, lacking effective implementation,[87] and since 2002, Ecuador is at the stage of elaboration of national access legislation.[88] Finally, Peru enacted the specific law on access to the biological diversity and the traditional knowledge in 2004.[89]

The objective of the regional effort with Decision 391 was to develop a harmonized policy in access and benefit-sharing for the Andean Community members. While domestic legislative action is necessary for the actual implementation of the Decision, it is observed that domestic measures taken by the countries in this case go beyond the Decision in that they run parallel to it and they diverge from the harmonized approach. For example, in Peru, Ecuador and Bolivia, there are two regulations, one applicable to scientific and commercial prospecting associated to the access to biological resources, and the other applicable to genetic resources. The same is valid for Ecuador and Bolivia. In Peru, the situation is even more complicated, because of the special legal regime on medicinal plants and on the promotion of the nutraceuticals.[90]

[85] Carrizosa in Carrizosa et al (n 1) 10; P Ferreira-Miani, 'Colombia: Access and Exchange of Genetic Resources', in Carrizosa et al (n 1) 79.

[86] Articles 15–17 of the Decree 309 of 2000, entered into force on 25/12/2000 ('Decreto 309 de 2000 por lo cual se reglamenta la investigación científica sobre diversidad biológica') address access to genetic resources and benefit-sharing for research purposes.

[87] Supreme Decree 24676 approving the regulation of the Decision 391 of the Cartagena Accord Commission and the regulation on biodiversity ('Decreto Supremo 24676—Aprueba reglamento de la decisión 391 de comisión del acuerdo de Cartagena y el reglamento sobre biodiversidad'), entered into force on 21/06/1997.

[88] Draft (12/042002) special law for the conservation and sustainable use of biodiversity in Ecuador ('Ley especial pata la conservación y uso sustentable de la biodiversidad en el Ecuador').

[89] See Law 28216 on the protection of the access to the Peruvian biological diversity and the collective knowledge of the indigenous peoples ('Ley 28216, Ley de protección al acceso a la Diversidad Biológica Peruana y los Conocimientos Colectivos de los Pueblos Indígenas'), entered into force on 01/05/2004. The Law 28216, articles 27–30 of the Law No 26839 on the conservation of and sustainable benefits from biological diversity, entered into force on 16/07/1997 ('Ley No. 26839 sobre la conservación y aprovechamiento sostenible de la diversidad biológica') and article 11and the Fourth transitory provision of the Supreme Decree No 068-2001-PCM that approves the Law on the conservation of and sustainable benefits from biological diversity, entered into force on 20/06/2001 ('Decreto Supremo no 068-2001-PCM Aprueban el reglamento de ley sobre la conservación y aprovechamiento sostenible de la diversidad biológica') were generally addressing access to genetic resources and benefit-sharing. Since 2001 Peru has a rather detailed national strategy on biological diversity (Estrategia Nacional sobre Diversidad Biológica) adopted on 01/01/2001 (strategic objective 2.8 Access to Genetic Resources).

[90] M Ruiz Muller, '¿Es necesarion un nuevo marco jurídico para, la bioprospeccion en la región Andina? Breve revision critica de la decisión 391' (Lima, Serie de Política y Devecho Ambiental, No 14 Sociedad Pervana di Derecho Ambiental, Programa de Asuntos Internacionales y Biodiversidad, 2003) 3.

The scope of Decision 391 encompasses traditional knowledge , but the issue is not yet comprehensively addressed. On 10 August 2002, Peru became the first country in the Andean region that adopted a national comprehensive legal system[91] for the protection of indigenous communities' collective knowledge associated with biodiversity.[92] Apart from the objectives set in Article 2 of the Decision (namely the establishment of conditions for just and equitable participation in the benefits of the access, promotion of conservation of the biological diversity, sustainable use of its components and promotion of the consolidation and development of scientific, technological and technical capacities in the region), there is the underlying objective of avoiding unfair competition among the Andean Member States with similar biological wealth. This objective derives from the purpose itself of the initiative for regional measures in access and benefit-sharing and is reflected in the duty of the national authorities to take into consideration the interests, viewpoints and information of the other Member States, when they negotiate the terms of access contracts to genetic resources that originated in more than one country.[93]

The Decision and the shortcomings of its implementation has been subject to criticism.[94] On 7 July 2002, the Andean Community adopted Decision 523 on the Regional Biodiversity Strategy.[95] This strategy includes an ABS component that describes some of the problems experienced by the Andean countries in implementing Decision 391 and proposes a course of action to facilitate its implementation.[96] Some elements of the Decision are further examined to evaluate the degree to which it meets the CBD objectives.

### 1. *Stakeholders' representation*

The stakeholders include: the State represented by the national authority, the owners of the biological resources (landowner, *ex situ* center, etc), the local and indigenous communities, the scientific institutions, and the industry.

[91] Law No 27811 establishing the regime of protection of the collective knowledge of the indigenous peoples associated to the biological resources, entered into force 10/08/2002 ('Ley que establece el régimen de protección de los conocimientos colectivos de los pueblos indígenas vinculados a los recursos biológicos').

[92] See also Venezuelan Law on Biological Diversity, Articles 84–91 ('Capitulo III de la protección y reconocimiento de los conocimientos tradicionales de los pueblos y comunidades indígenas y locales').

[93] Decision 391 Second Final Provision.

[94] M Ruiz Muller (n 90) 1.

[95] Decisión 523 Estrategia Regional de Biodiversidad para los Países del Trópico Andino.

[96] Carrizosa in Carrizosa et al (eds) (2004 n 1) 10.

a) The State

Decision 391 affirms the CBD principle of sovereignty. Furthermore, the issue of the legal status of genetic resources is handled in Article 6 of the Decision 391, but notably, the Decision's provision is unclear on the precise legal status of the genetic resources: genetic resources and their by-products are either patrimony of the Nation or property of the State.

In the former instance of national patrimony, the degree of private participation and freedom to negotiate an access procedure is greater than the latter of national property, where the State has the right to become the exclusive provider of genetic resources. If genetic resources are patrimony of the Nation, the State physically administers them in favor of all citizens and does not necessarily have an exclusive right to negotiate access agreements; negotiable rights, short of full property rights, over elements of national patrimony may, under certain circumstances, be granted under law by way of concession or otherwise. If genetic resources are the property of the State, this legitimizes the State's right to negotiate access agreements.[97]

The role of the State with respect to genetic resources represents the ideological and conceptual differences between civil law and common law. For historical reasons, the Andean countries embrace the Roman legal tradition, where natural resources, in general, are part of the national patrimony or property and therefore the State has a pivotal role in regulating access to and use of them. Common law countries, on the other hand, are primarily based on private property rights and thus limit the role and participation of the State in regulating access to and use of natural resources.[98]

As a result, in either case of national patrimony or of State property, the role of the State remains important, but the degree of State intervention and involvement in the negotiations could vary from the negotiation of access agreements to supervision of the agreements, establishment of minimum conditions, provision of permits or authorization according to national legislation. This has serious consequences in the national transposition of the Member States.[99]

The Member Countries of the Andean Community share the Roman civil law tradition, but have different views on State involvement, which is reflected in their Constitutions.[100] During the negotiating process, Member

97 Ruiz in Mugabe et al (eds) (n 1) 192–193.

98 Ruiz Muller (n 1) 4–5.

99 ibid.

100 In Colombia and Ecuador, genetic resources and their by-products are the goods or property of the nation, ie they 'belong' to all the inhabitants of these countries. The State, acting as a representative of the inhabitants' interests, is empowered to manage and control the use of those genetic resources. For this reason, Colombia sought recognition of the State alone to negotiate access agreements and did not accept that indigenous people may negotiate on

Countries could not come up with a compromise. This resulted in Article 6 of the Decision, which is broad enough to allow for the tailoring of the new access regime to the traditions and political and social contingencies of each country.

Thus, genetic resources in the Andean common regime are inalienable and not subject to prescription being considered either national patrimony or State property. Contrary to genetic resources, biological resources remain subject to private property rights. The Decision is innovative in that it defines genetic resources and their by-products as self-contained objects of legal protection.

According to this legal device, the access conditions to genetic resources are stipulated in the contract between the applicant and the State (access contract). The access and use conditions to the biological resources containing the sought genetic resources are further negotiated and agreed upon with the owner/manager of the land and/or of the biological resources (ancillary contracts), be it private persons, entities, or indigenous peoples. The suppliers of the biological resources have no rights over genetic resources and their by-products.[101]

b) Local/indigenous communities

Among its definitions in Article 1, Decision 391 makes clear reference to the supplier of the intangible component, and it further requires that national legislation should provide for a strict protection of indigenous peoples' interests and rights.[102] It is indeed in the interest of the State to make official contact with the indigenous communities in its territory. The Andean Framework though fails to protect indigenous rights. The rights of indigenous communities over their lands are not considered, and their collective rights to knowledge, although pronounced, are insufficiently protected.

The Decision refers only to the intellectual contribution of indigenous peoples implying that the State is the sole owner/administrator of indigenous lands. Indigenous peoples' rights over material resources are not acknowledged and thus indigenous peoples are not entitled to ancillary contracts for the biological resources/lands they manage. Indigenous peoples have rights to use land/resources for subsistence purposes, but these rights do not culminate in the right to negotiate and conclude contracts for the purposes and interests concerning bioprospecting activities or to claim

biological resources in side agreements without the participation of the State. In Peru genetic resources are the patrimony of the nation and their protection and conservation is part of the social interest. Hence, during the negotiations, Peru sought to limit the participation of the State in private sector activities and the State control over the process; see Chaves in Stoianoff (ed) 2004 (n 55) 244 and Ruiz in Mugabe et al (eds) (n 1) 197.

101 Decision 391 Articles 41–44; see Section 3.2b).

102 ibid Article 7.

benefits deriving from the use of the genetic resources contained in the resources and lands they manage.

Land rights and interests of the indigenous peoples could possibly be justified by Article 34 of Decision 391, which reads that the 'access contract shall bear in mind the rights and interests of the suppliers of genetic resources and their by-products, the biological resources that contain them and the intangible component'. But this wording is vague and does not signify the actual involvement of the indigenous populations in the negotiation of the contracts and the benefits.

With regard to intangible contribution of indigenous peoples, Decision 391 appears to go further than CBD Article 8j; its Article 7 reads that

> The Member Countries, in keeping with this Decision and their complementary national legislation, recognize and value the rights and the authority of the native, Afro-American and local communities *to decide* about their know-how, innovations and traditional practices associated with genetic resources and their by-products (emphasis added).

Decision 391 does not specifically stipulate how the contract concluded between indigenous peoples supplying the 'intangible component' and the applicant/prospector would meet the fair and equitable objectives.[103] This is then left to the Member Countries, and protection of indigenous peoples' rights varies depending on the national indigenous peoples' policy. Member States were expected to present their national studies by 2000, on this basis a harmonized regime would be designed at the regional level in support of the protection of 'intangible component' of native, Afro-American and local communities.[104] These studies have yet not been presented. As a result, the recognition of the rights of indigenous/local/ Afro-American communities over their traditional knowledge does not have any practical expression or enforcement.

In fact, the negotiations phase of the Decision aimed at a stronger protection for the indigenous rights, with a series of measures for the control, development and protection of the cultural expressions and the biological and genetic resources, but ultimately this line of proposals did not pass. At the non-governmental phase of the process, the concept of 'cultural objection'[105] was discussed as a possible justification to prevent prospecting activities which indigenous communities found were not in their interest.

[103] Decision 391 Article 34. [104] ibid Eighth Temporary Provision.

[105] The right to cultural objection is stipulated in Article 66 of the Decree 7788 Biodiversity Law of Costa Rica, according to which local communities and indigenous peoples have the right 'to oppose any access to their resources and associated knowledge, be it for cultural, spiritual, social, economic or other motives—unofficial translation by Berhard Mucahy with the assistance of GRAIN in June 1999). See also S Pena-Neira (n 1).

Also, a special regime was proposed to ensure that the will of the communities was effectively respected; 'cultural consultation' and PIC would be necessary means to ensure that communities freely decide whether they want to enter in negotiations or to contact potential applicants. But, even this suggested enhanced protection did not include the clear recognition of specific rights over resources.[106]

According to the Decision, Member States as representatives of their citizens' interests have the right to prohibit access to genetic resources when this may result in violations of cultural or environmental rights.[107] However, since the development of the environmental and social impact assessments are not compulsory, it is difficult to envisage how States will be able to detect such violations. On the whole, indigenous peoples' rights to the tangible and intangible components are rather weak and absence of a PIC procedure is the main drawback of the Decision in this respect.

c) Scientific community

Decision 391 creates a favourable framework for the promotion of research and technology in the territory of the country where the access takes place, and the supply of information regarding science and background of the resource to be accessed. However, domestic scientific institutions were neither consulted during the negotiating stage nor are effectively granted an important role in the access process. The deposit/administration/intermediation contracts, where the scientific institutions are parties[108] are not compulsory and cannot justify the due participation of the universities/research institutions/researchers.

d) Industry

There is no indication that industry has been consulted at the time of discussion of the Decision 391. If the access and benefit-sharing regulation is geared up to maximization of income through the commercialization of genetic resources and the channeling of the profits to conservation and sustainability purposes, industry should be consulted, as well as research institutions.

e) NGO community

A very dynamic NGO community characterizes the region. Also, at the preparation stage, as noted above, effort was put to include wide participation by NGOs, but because of disagreements, governments eventually decided to abandon this approach and exclude NGOs from the process.

[106] Chaves in Stoianoff (ed) (n 55) 245–246; Ruiz in Mugabe et al (n 1) 194.
[107] Decision 391 Article 45 lit. c) and also the task of the Autoridad in Article 50 lit. d).
[108] Decision 391 Fifth Complementary Provision.

### 2. Application Procedures for Agreements and Obtaining PIC

The role of the State and the duality of rights with regards to genetic and biological resources in the Andean common access regime created a complex contractual system, which involves the following contracts:

- *Access contract:* to the genetic resources—between the State and the applicant;[109]
- *Ancillary contracts:* between the State and the owner/possessor/manager of the land where the biological resources containing the genetic resources are located and/or the *ex situ* center and/or the owner/possessor/manager of the biological resources containing the genetic resources and/or the national support institution;[110]
- *Deposit contracts, Administration contracts and Intermediation contracts:* between the State and universities, research centres and researchers;[111]
- *General framework agreement:* between the State and universities, research centres and researchers;[112]
- *Annex contract:* for the use of the intangible component.[113]

In reality, it is unusual that in a given situation all these types of contracts exist simultaneously, but there are at least two of the above forms present each time. The whole procedure may lead to a complicated, burdensome and costly processing for the potential user.

Combined with the inadequate protection of indigenous peoples' rights, the absence of the PIC process in the Decision is a considerable shortfall. The contract signed for the intangible contribution of the indigenous peoples is actually incorporated in the main access contract where the main party is the State; the State actually negotiates the royalties for their supply.[114] Alternatively to this contractual matrix, States would better form a joint negotiating position together with providers in general, and indigenous peoples in particular, protecting individual and collective rights.

### 3. Sharing the benefits

The Decision's focus is on access and less on benefit-sharing. As such, it fails to establish the conditions to make its objective of fair and equitable benefit-sharing[115] operational, because it is vague on the type of the benefits, the beneficiaries, and the distribution mechanisms. It leaves everything to the

[109] ibid Article 32.
[110] ibid Article 41.
[111] ibid Fifth Complementary Provision.
[112] ibid Article 36.
[113] ibid Article 35.
[114] ibid.
[115] Decision 391 Article 2a).

discretion of the Member States to decide. Application requirements set by the Decision (such as the culturally, socially and environmentally sustainable and safe transfer of knowledge and technology, and the in-country improvement of scientific knowledge on the status of biological diversity)[116] in as much as the establishment and financing of funds and other financial mechanisms with the profits arising from the access with the goal to promote compliance with the aims of the Decision, and hence with the aims of the CBD,[117] lack accuracy and effectiveness. In the end it is the State, as the main actor, that most probably will profit from the commercialization of genetic resources, but it is not at all evident how it will forward the shares to the rest of the stakeholders.

With regard to indigenous peoples, the member countries are asked to design a capacity building training programme,[118] but this is not obligatory for the States and it does not guarantee the indigenous peoples' rights. Indeed, the design of a mechanism for the fair and equitable sharing of the benefits from the use of genetic resources or intangible components related to indigenous peoples, while providing them with effective protection for their collective rights, is a very complex task. Lacking a domestic legislation, the affirmation of the indigenous peoples' tangible component (the use of genetic resources) is a more difficult task than the identification of their intangible component; despite the controversy surrounding traditional knowledge, and irrespective of the uncertainties over the individual or collective nature of the right, it is easier to identify that traditional knowledge belong to somebody.

### C. *Final Remarks on the Andean Pact*

Commentators have both exalted and criticized the first comprehensive regional attempt on ABS regulation.[119] In spite of its environmental and conservationist character, Decision 391 primarily serves the commercial aim to regulate access to and trade of genetic resources, without even consulting with industry sector. In fact, in a world governed by economic interests, a purely conservationist blueprint would completely fail to meet the CBD objectives. At the same time the pressure of the international market for natural resources may heavily influence biodiversity rich countries in exercising sovereign rights over their genetic resources, because they overestimate the option values of their resources.[120]

A well designed regional framework significantly strengthens the negotiating position of the member countries against industrialized countries and

116 ibid Article 17 lit. c) and d).
117 ibid First Complementary Provision.
118 ibid Ninth Temporary Provision.
119 Chaves in Stoianoff (ed) (n 55) and Ruiz (n 90).
120 Chaves in Stoianoff ibid 239.

biotechnology industries,[121] as it offers opportunities for cooperation and it keeps away from anti-competitive practices among the countries sharing natural resources, ecosystems and environmental characteristics. But the deficiencies of the Decision offset this potential advantage.

Decision 391 has a much broader scope than the CBD, covering genetic resources, their by-products, and the intangible components associated to the resources or the by-products. More specifically with regard to traditional knowledge, the Decision covers all knowledge of actual or potential value, communities' and individual knowledge, knowledge associated to genetic resources, by-products and the biological resources containing them, knowledge important for the conservation and sustainable use of biological diversity, but all the knowledge of actual or potential value, in whole the scope goes beyond the CBD Article 8j.

Even though indigenous knowledge seems to be more extensively protected, the absence of PIC and the State being 'behind' the annexed contract for this knowledge, speak for the inadequate transposition of the CBD prerogatives. By distinguishing the ownership over biological material from the ownership over genetic and/or biochemical resources, Decision 391 comes up with an innovative ownership construction: it creates a new object of legal protection, the genetic material, which is awarded different legal treatment in relation to the ownership status of the biological 'bearer' of the material. Genetic resources are declared as State property or heritage; they belong to the public domain, and whether found on the beach or on the top of a mountain, they are never free of disposal. This does not cause any difficulties in the private ownership and general disposal rights of the first material. The applicant for access is requested to obtain the consent of, and to enter into ancillary agreement with, the owner of the land/biological resources containing the genetic resources. The State does not give up the access to the private ownership of the primary 'bearer' material, but it binds it to the access contact.

The problem here is that the duality of rights splits the access procedure by bringing about a complicated contract procedure, involving many bureaucratic acts and economic costs, which may discourage potential applicants.

A drawback of the Decision is that it does not differentiate between research and commercial use, and it submits the access to genetic resources for academic purposes in the same extremely complex process. There is the

[121] eg access to genetic resources was a point of discord for several months in 2004 in the free trade agreement that the US had been negotiating with Colombia, Ecuador and Peru; see S Surrii, Potato Capital of the World Offers up New Recipe, Inter Press News Agency, London, January 18, 2005, published on the Internet at http://www.ipsnews.net/ interna.asp?idnews=27069 (last seen 30/06/09) and C Vieira-Tierramérica, 'Environment: Andean Nations Clash with U.S. on Patenting Life Forms', Inter Press Service News Agency, Bogota, November 2004.

reasonable argument that it is very difficult to draw the limits between academic research and commercial use, however, an easier administrative path should be sustained for domestic institutions. The content of the obligations, the technology transfer and participation in research, and generally the benefit-sharing provisions of the Decision are inadequate in themselves. Specifically with regard to technology transfer, biotechnology and environmental provisions, the Decision actually refers to national regulations; for instance, it merely forbids the use of biological weapons and practices that harm the environment or human health,[122] without further defining them.

With so many broadly defined obligations, the Decision is asking too much of the national legislation and the national institutions. On the one hand, it provides flexibility necessary for the Member Countries to accommodate their local particularities, but on the other hand the degree of its effectiveness, especially when one considers the steps taken by the Member Countries so far is questionable. If the decision repeats the CBD principles and leaves the Member Countries to their own access and benefit-sharing devices, the purpose of a regional legislation is certainly not achieved.

To remedy its drawbacks, a proposal for the Decision's revision has been underway since 2001.[123]

## IV. SUMMARY AND CONCLUSION

The main challenge with domestic implementation is to what extent should the State interfere in the regulation of the bioprospecting market with resources originating from its territory. The prevailing role of the State and the complexity of the process reveal that source countries have high expectations for genetic resources, designing policies and legislations that attempt strictly to control genetic resources and biologically derived material.

The definition of property rights is critical for a safe contractual environment. Clear property rights over lands and resources determine the contracting parties to the bioprospecting agreements. The Andean Decision's duality of rights over the material creates a contractual situation, where there is a risk that the interests of the stakeholders, in particular those of the indigenous peoples, are overseen. Both the Philippines and the Andean Pact countries subject *ex situ* facilities to the same procedures as *in situ* resources. In light of the interdependence of the countries all over the world, for access to genetic resources from other States and food security

[122] Decision 391 Articles 24 and 25.

[123] See the Propuesta de Reglamento sobre Acceso a los Recursos Genéticos, Abril-Julio 2001, on the part of the Instituto Nacional de Recursos Nacionales (INRENA), Ministerio de Agricultura, Peru.

claims, access to *ex situ* resources should be loosened to prevent unnecessary delays and ensure food supplies.

Another point is that sometimes additional regulations are applied for resources located in protected areas,[124] and the applicant may be called to comply with several layers of approval: the regular access and benefit-sharing law, the national legislation related to research activities in protected areas, and eventually CITES.[125]

Indigenous rights are protected by a separate law in the Philippines, the Indigenous Peoples' Rights Act. Such a legal framework is absent in the Andean Community. In fact, vast indigenous populations inhabit Andean countries, manage the resources, and develop medicinal knowledge. The failure of the common regime effectively to involve indigenous communities in the negotiations and the benefits is inexplicable.

Regional approaches are a useful response in ABS legislation for neighboring countries with cross-border genetic resources. First, regional frameworks enhance cooperation at political and technical level; States with geographical and ethnic similarities benefit from a common ABS regime, because they can easily exchange information, technological capacity and expertise; and they do not compete with each other on prejudicial terms, while they have a stronger voice at the international negotiating arena. Secondly, streamlined processes provide greater predictability for users and promote investment and development in the region. Unfortunately, the Andean Decision missed the expectations as it did not manage to provide for regional legal certainty, and there are many parallel ABS measures in the Member Countries; only secondary details should be left to the Member Countries, while a common core is necessary.

[124] Eg Decision 391 Sixth Complementary Disposition: 'when access is requested for genetic resources or its derived products from protected areas, the applicant will comply with provisions of Decision 391 and with specific national regulations on the matter'.

[125] Peru offers an interesting case of increasingly complex legal structures in protected areas. Through Supreme Decree 010-99-AG of 7 April 1999, the Director Plan for Natural Protected Areas was approved, laying out the policy and regulatory framework for the management and administration of different types of protected areas in the country. Among other, the Decree establishes specific requirements for general scientific research in protected areas and refers to biodiversity prospecting (Point 2.3 of Chapter II). Access to genetic resources is subject to the objectives of the protected area, the land-use plan and the management plan for the particular area. These conditions run in addition to those established by Decision 391; but the Authority responsible for granting the research permits in protected areas (INRENA) is different from the authority for granting permits for access. See also Ruiz in Laird (ed) (n 12) 172.

# CHAPTER 6

# *Private Practice in Bioprospecting*

## I. INTRODUCTION

Private policies on access and benefit-sharing communicate the views of researchers, institutions and companies on the ethical and practical issues associated with the use of biodiversity. They evolve through practice and have the potential to steer policy-making. This chapter examines how institutional and corporate policies and professional guidelines shape access and benefit-sharing in drug industry.

The bilateral system refers to direct agreements between suppliers and users of genetic resources. The CBD favours the bilateral system of exchange of genetic material in contrast to the multilateral system established by the FAO International Treaty for the acquisition and exchange of genetic resources for food and agriculture.[1] These contracts are commonly known as Material Transfer Agreements (MTAs), and, depending on the prospective use of the transferred material, they are divided in academic research agreements and more complex commercial research agreements.[2]

Research agreements aim at describing the material, and they include a statement that commercialization would require another agreement, while distribution to third parties is prohibited.[3] Commercial agreements need to fulfil more requirements and there are many actors representing sometimes conflicting interests. Apart from the user and the source country, there is usually a research/academic institution involved either in the user country or in the source country or in both, who play the role of intermediaries, and also those who may have contributed to the identification of a commercial product, typically a local/indigenous community. There is the option of one multiparty complex agreement or the option of several bilateral agreements coordinated by a hub institution.[4]

The supply of genetic resources is conditioned on the permission of their owner. Problems in MTAs arise when the owner of the resources is not iden-

[1] See Chapter 4.

[2] W Lesser, *Genetic Resources—Convention on Biological Diversity: Exploring Access and Benefit-Sharing Issues* (CAB International, Oxon, New York, 1998) 29 f.

[3] ibid 29–31.

[4] ibid 32–40; see also the Model Biodiversity Prospecting Agreement by D Downes et al, 'Annex 2: Draft Biodiversity Prospecting Contract' in WV Reid et al, *Biodiversity Prospecting* (World Resources Institute, Washington DC, 1993) 255 f, which is more like a Code of Conduct for the collector/user of the material.

tifiable. Domestic access and benefit-sharing legislation is expected to provide for the MTA's general terms and conditions. Presently only a few countries dispose of access and benefit-sharing legislation; most countries are still at the design stage.[5]

In response to the absence of national and international rules for the collection and exchange of genetic resources, the growing awareness raised during the last decade motivated many public and private actors in biological research and prospecting to develop and promote their own policies in access and benefit-sharing. A range of institutional policies, corporate policies and professional societies standards evolved that take into account the social and political aspects of their activities raised by the CBD. These policies show the users' or collectors' commitment to the CBD, but they should by no means be considered to replace domestic regulation on access and benefit-sharing; their function is of a complementary character.

In the framework of its Global System on Plant Genetic Resources, the FAO prepared codes of conduct and international standards. In 1991 the FAO issued the Draft International Code of Conduct for Plant Germplasm Collecting and Transfer, adopted by the FAO Conference in 1993. The FAO code of conduct is intended to provide guidance to national governments until they implement national access and benefit-sharing measures. The code is in full accord with the CBD because it secures implementation of national sovereign rights, and can be seen as a model law for implementing countries, but it completely depends upon the will of the users and/or collectors of genetic material.[6]

The 'standard terms' for genetic materials have yet to arise and this is a great challenge to suppliers of genetic resources. The most difficult matter is perhaps the identification of the objectives/expectations of countries and communities for their genetic resources. The efforts to set a standard MTA in the field of plant genetic resources has proved already a very difficult task,[7] and it is much more difficult to set such a standard agreement covering all sorts of genetic resources on the premises of the CBD.

[5] See Chapter 5.

[6] See International Code of Conduct for Plant Germplasm Collecting and Transfer, adopted by the FAO Conference in 1993. Also, a Draft Code of Conduct on Biotechnology has been prepared in 1993 under the auspices of FAO, with the objective to 'maximize the positive effects and minimize the negative effects of biotechnology' (see Commission on Plant Genetic Resources, 'Towards an International Code of Conduct for Plant Biotechnology as it affects the Conservation and Utilization of Plant Genetic Resources', Doc CPGR/93/9, February 1992). The discussions of the draft were subsequently postponed in view of the CBD Biosafety Protocol and the revision of the FAO International Undertaking.

[7] A Smagadi, 'Accès aux ressources génétiques et partage des avantages découlant de leur exploitation: la Convention sur la diversité biologique et le système mondial de la FAO' (2005) 3 Révue Européenne du Droit de l'environnement 255.

## II. INSTITUTIONAL POLICIES

Research institutions involved in biodiversity research and prospecting can be universities, botanic gardens, natural history museums, government research institutes and other non-profit institutions that receive some level of support from the government. During the last decade, a very great number of these institutions developed their bioprospecting policies.[8] As biodiversity research and prospecting and the exchange of materials in *ex situ* collections evolves, the reason behind this rapid development of institutional bioprospecting policies worldwide is due to the growing need for clarity and transparency of institutional activities and policies in the absence of international and domestic policy context; institutions active in research activities are called to formalize their policies on issues associated with material collection, prior informed consent, benefit-sharing, and institutional commitment to pertinent international treaties, such as the CBD and CITES.[9]

The development process of these policies is a hard exercise that requires considerable awareness-raising and capacity building. It is based on continuous follow-up strategies, refinement, and adaptation to the lessons learned from this complex international dialogue underway on access and benefit-sharing. Individuals working for the institution or external collaborators may be reluctant to comply with bureaucratic and restrictive policies. Some institutions develop an instrument to specifically address commercial acquisition of natural products, and other develop a more comprehensive policy covering all benefit-sharing aspects of both research and commercial activities.[10]

On assessing these two policy options, the same arguments apply as in the design of access and benefit-sharing legislation.[11] Research on biodiversity deals with collection, survey, cataloguing and assessment, activities

[8] See the Table 3.1 detailing institutional biodiversity prospecting policies by SA Laird and R Wynberg, 'Institutional Policies for Biodiversity Research' in SA Laird (ed) *Biodiversity and Traditional Knowledge—Equitable Partnerships in Practice, People and Plants Conservation Series* (Earthscan Publications, London, 2002) 42.

[9] ibid 39 f.

[10] ibid 41 f.

[11] See Chapter 5 on the Philippines: EO 247 Section 3 distinguished between Academic Research Agreements (ARAs) and Commercial Research Agreements (CRAs), but whenever the user was a foreign entity, the agreement fell by definition under the stricter CRAs rules, because it would be difficult to track the future uses of the provided material, even if its first and declared use was only for academic purposes. The same criterion of domestic entity involved for the characterization of the agreement as ARA applies also with the new Wildlife Act and its Implementing Rules that replaced EO 247; for the ARA the execution of an Affidavit of Undertaking or Memorandum of Agreement by the applicant and the issuance of a gratuitous permit by proper or concerned authorities are necessary (Wildlife Act Section 15 and Implementing Rules), whereas stringent requirements apply for the Bioprospecting Undertaking, as the previously known CRA is now called.

for which permission should be granted under looser conditions, if the dividing line between academic and commercial research were not as blurred as it is. The fact is that academic research is regularly coupled with commercial applications, and the commercial partnerships between research institutions and drug companies are numerous.[12]

A group of 17 botanic gardens has made the effort to harmonize their practices in access to genetic resources and benefit-sharing by developing the Common Policy Guidelines for Participating Botanic Gardens on Access to Genetic Resources and Benefit-Sharing (2001). The Common Policy takes on the general best practices and standards in collecting and exchanging genetic material, which should be adapted to the concrete cases and institutions each time, and it is considered to be a great exercise for people working for botanic gardens, because they are now obliged to take into account social and political issues raised by biodiversity research in conjunction with their biological and ecological tasks.[13]

### *A. The National Cancer Institute and the Calanolides*

The National Cancer Institute is one of the seventeen US National Institutes of Health (NIH) under the auspices of the Department of Health and Human Services of the US Government. It was established in 1937 with the mission 'to provide for, foster and aid in coordinating research related to cancer'; in addition to the search for the anti-tumor agents, the NCI initiated a programme to find natural products for the treatment of AIDS in 1988.[14] As a government funded, non-profit organization, the NCI's mandate does not allow it to engage in the commercialization of any of its products, but may participate in the drug development process until the stage of commercialization.

The NCI has demonstrated a strong interest in natural products all over the world, and focused its research especially in the tropical regions (but plants are collected also in Africa and Asia) for the investigation of the potential for the development of novel medicines based on the known properties of plants, marine organisms and microbial cultivations.

In the framework of its collaborations, NCI has developed access and benefit-sharing policies. While in the past a permit from the country where research was taking place was sufficient, the 'biopiracy' discourse demanded the elaboration of policies where the rights of the country providing the material would be better protected in case of commercializa-

[12] ibid 54, see Table 3.2 on Examples of research institution commercial partnerships in the 1990s.

[13] ibid 50 f.

[14] www.cancer.gov.

tion. As these policies evolved, we observe significant changes in the benefit-sharing terms of the agreements: each one places greater emphasis than its predecessor on technology transfer, collaborative research, and value to be added in the source country.[15]

Two studies about the NCI policies, on the Calanolides and Topotecan,[16] were submitted to the CBD Secretariat for consideration at the COP4. While it is some time ago that these were submitted, the Calanolides case is here described to highlight the successful joint venture in the country providing the material and also the difficulties envisaged in federal states.

The anti-HIV Calanolides compounds (*Calanolide A* and *Calanolide B*) are isolated from two Calophyllum species that occur naturally in the rainforests of Sarawak state, Malaysia. The stakeholders involved in this undertaking were the NCI, the State Government of Sarawak, the private biotech company Medichem Research, and Sarawak-Medichem Pharmaceuticals, a joint venture formed by the Government of Sarawak and Medichem Incorporated. Other actors were the State Forestry Department in Sarawak and the University of Illinois at Chicago (UIC).

This case shows how complex regulation may be in federal states, creating a negative investment climate, but also how circumstances may trigger legislature initiatives. To collect plant species, a research permit was (and is) required at the federal level for any foreigner who wishes to do so. At the state level, states like Sabah and Sarawak, where the Calophyllum were collected, have their own laws on forests and wildlife protection and require additional permission to be issued for the purposes of plant collection. Further to its experience with the calanolides that lead to the establishment of a joint venture in the state, Sarawak state soon amended its forest legislation to include access and benefit-sharing provisions.

Interestingly, not only countries get motivated to adopt or amend their ABS policy after they experience bioprospecting practice, but so do private companies. For instance, the NCI Topotecan case was the impetus for Smith Kline Beecham to change its policy and improve its corporate image in this respect.[17]

From a structural point of view, there were three separate agreements connecting the main actors: a Letter of Collection between the NCI and the State Government of Sarawak; a Cooperative Research and Development Agreement (CRADA) between NCI and Medichem Research

[15] K Ten Kate and A Wells, 'The Access and Benefit-sharing Policies of the United States National Cancer Institute—A Comparative Account of the Discovery and Development of the Drugs Calanolide and Topotecan' Submission to the Executive Secretary of the Convention on Biological Diversity by the Royal Botanic Gardens, Kew, 1998.

[16] ibid.

[17] ibid 28.

Pharmaceuticals, and an agreement between Medichem Research and the State Government of Sarawak establishing Medichem Sarawak Pharmaceuticals.

The relationship between the NCI and the Sarawak State was formalized after a couple of years of initial collections in 1994 with 'Agreement between the Sarawak Forestry Department and the Developmental Therapeutics Program of the Division of Cancer Treatment (NCI)' (Letter of Collection). Meanwhile in 1992, the NCI had filed patent applications for Calanolides, and in 1993 it awarded a grant to Medichem Research Incorporated, whereas a couple of years later NCI awarded Medichem a worldwide, exclusive license to its patents on the preparation and use of Calanolides and it also required that Medichem negotiate an agreement with the Sarawak Government, thus fulfilling the NCI's obligations under the 1994 Letter of Collection. The joint venture was officially established in 1997[18] and the Medichem acquired exclusive worldwide licenses now belonging to the joint venture, with the state of Sarawak being a joint licensee.

The establishment of a joint venture between a private pharmaceutical partner and the state of Sarawak in Malaysia was a novelty. The state provider of the genetic resources receives not only monetary benefits from the commercialization of the collected resources, but it has the great opportunity for training in scientific and business practices. The key goals of the joint venture relate to the future of the Calanolides, the investigation of other drug candidates in the rainforests of Malaysia (as part of Sarawak's approach to sustainable development), and the training of Malaysian scientists.

The significance of NCI policies is greater considering that the NCI is a US public sector institution, based in the US and financed by the US government funds, and that the US is not a party to the CBD and is actually one of the main players at the international level objecting to an access and benefit-sharing instrument and to other proposals, such as PIC certification and disclosure of origin, which aim at facilitating the international ABS regime. It is also very important that the State institution binds in its agreements the private sector to build partnerships promoting the economic and environmental interests of the source country.

The Calanolides case was definitely not very difficult to develop, because neither indigenous peoples nor traditional knowledge were involved. The resources were found in the forests and there was clarity with regard to the property rights over the genetic resources. In any case, sustainable sourcing should be a priority in the bioprospecting policies and governments should pay particular attention to the monitoring of the collection practices, and to the quality and the quantities of samples collected. Companies, even if they

[18] www.sarawak-Medichem.com

synthesize products, usually need to obtain larger quantities of raw plant materials during the discovery and development phase. The Calanolides and the Taxol cases are indicative of species loss because of imprudent collection.

### *B. The International Cooperative Biodiversity Groups Program in Peru*

The International Cooperative Biodiversity Groups (ICBG) Program, established in 1992, is an experimental program, funded by the United States Government. The objective the Program is manifold: to improve human health through the discovery of natural products with medicinal properties to treat diseases of importance for both developed and developing countries; to conserve biodiversity through the valuation of natural resources and to develop local capacity to manage these natural resources, by creating incentives and increase knowledge; and to promote economic growth by sharing the benefits of the drug discovery and conservation research processes.[19]

The Groups are financed through cooperative agreements. Each of them is a consortium of academic institutions, local and international private voluntary organizations, and in most cases a private pharmaceutical company. One or more of the partner organizations is based in the source country. The coordination is carried out by an academic principal investigator, who directs his or her own program in natural products chemistry, drug development, or ethnobiology. Training and other local capacity building efforts are usually part of each of the associate programs. The formation of the groups illustrates the international cooperation called for by the CBD in Articles 5, 10(e), 18.5.

ICBG Projects are approved as long they meet the following criteria: full disclosure and informed consent, sharing of both near- and long-term benefits with the source country communities and organizations, respect of the pertinent international and local laws, and that credit is given to local indigenous or other intellectual contributors whenever possible. The

[19] JP Rosenthal (1998) Activities of the International Cooperative Biodiversity Groups program, The International Cooperative Biodiversity Groups Program, A Benefit-Sharing case study for the Conference of Parties to Convention on Biological Diversity, submission to the COP 4 of the CBD; J Rosenthal, 'Integrating Drug Discovery, Biodiversity Conservation and Economic Development: Early Lessons from the International Cooperative Biodiversity Groups' in F Grifo and J Rosenthal (eds) *Biodiversity and Human Health* (Island Press, Washington DC and Corelo, California, 1997) 281 f; F Grifo, 'The Role of Chemical Prospecting in Sustainable Development: An Overview of the International Cooperative Biodiversity Groups Program' in J Feinsilver (ed) *Emerging Connections Among Biodiversity, Biotechnology, and Sustainable Development in Health and Agriculture* (Pan American Health Organization, Washington, DC, 1996) 560 f; J Schweitzer et al, 'Summary of the Workshop on Drug Development, Biological Diversity, and Economic Growth' (1991) 83 Journal of the National Cancer Institute 1294; also ICBG web site: www.icbg.org.

programme funds projects in a number of countries.[20] In the following sections the project 'Sustainable Management of Medicinal Plants and Ecosystem Conservation in Northern Peru' is examined.[21] The area where the material was collected was located in Northern Peru, inhabited by five ethno-linguistic groups, of the Jívaro linguistic family. Among the other ICBG projects, it is particularly interesting for the current study, primarily because of the form of traditional knowledge protection eventually adopted with the second set of negotiations, ie a know-how licence between the Collaborating Organizations and the private party, the pharmaceutical company Monsanto-Searle. Further, the case indicates the problems of representation for indigenous peoples in the negotiation of the agreements.

The partners involved in this undertaking were Washington University (WU) which was coordinating the project; two Peruvian Universities, the Universidad Peruana Cayetano Heredia (UPCH) and the Universidad San Marcos; the Aguaruna peoples (organizationally represented by Consejo Aguaruna-Huambisa-CAH); and finally, in the framework of the ICBG program's active encouragement of the private sector collaboration, WU secured the participation of Searle & Co.

The contract design started in a legal vacuum with no international or national law to govern it. At the international level, the CBD came only into

[20] D Soejarto et al, 'Implementing IPR and Benefit-sharing arrangements: Experiences in the University of Illinois at Chicago-Vietnam-Laos ICBG' in JBA/NITE, *Intellectual Property Rights and Traditional Knowledge on Genetic Resources in Pharmaceutical and Cosmetic Business* (JBA/NITE, Tokyo, 2002) 47 [Proceedings of a Symposium organized by Japan Bioindustry Association (JBA) and National Institute of Technology and Evaluation, Tokyo, Japan]; and D Soejarto et al, 'An International collaborative program to discover new drugs from tropical biodiversity of Vietnam and Laos' (2002) 8 Natural Product Sciences 1; M Guérin-McManus et al, Bioprospecting in Practice: A Case Study of the Suriname ICBG Project and Benefits Sharing under the Convention on Biological Diversity, 1998, Case study submitted to the CBD Secretariat; M Iwu and S Laird, The International Cooperative Biodiversity Group—Drug Development and Biodiversity Conservation in Africa: Case Study of A Benefit-Sharing Plan, Case Study submitted in response to the Call for Benefit-sharing Case Studies by the Secretariat of the Convention on Biological Diversity, 1998.

[21] See W Lewis et al, 'Advantage of ethnobotanically based research when searching for new pharmaceuticals' (2004) 16 Ethnobotany 1; W Lewis and V Ramani, 'Ethics and Practice in Ethnobiology: Analysis of the International Cooperative Biodiversity Group Project in Peru' (2003) *Index to Conference Papers and Commentary, Biodiversity and Biotechnology and the Protection of Traditional Knowledge*, 4–6 April 2003; B Tobin et al, 'The Lessons of the Peruvian ICBG Agreements', paper presented at the Roundtable *User Measures and International ABS Governance* (Paris, Institut du Développement Durable et des Relations Internationales (IDDRI), Paris 6–7 November 2003) and personal communication with Brendan Tobin, legal advisor of the collaborating organizations with the Sociedad Peruana De Derecho Ambiental (SPDA), then research fellow at the United Nations University/Institute of Advanced Studies, Tokyo, Japan; also, S Greene, 'Indigenous Peoples Incorporated? Culture as Politics, Culture as Property in Contemporary Bioprospecting Deals' (2004) 45 Current Anthropology 211; and S Greene, 'Intellectual Property, Resources or Territory? Reframing the Debate over Indigenous Rights, Traditional Knowledge, and Pharmaceutical Bioprospection' in M Bradley and P Petro (eds) *Truth Claims* (Rutgers University Press, New Jersey, 2002) 229 f.

force in the end of 1993, but in 1994, when the Peru ICBG planning started, it was still very early to come up with interpretations of the vague and broad relevant CBD provisions. Peru is one of the developing countries that played an active role in international debates on ABS during negotiation of the CBD and regionally in debates relating to the development of a regime for protection of rights over plant varieties, at the level of the Andean Pact.

The Regional Legislation (Decisions of the Andean Community) involves the (i) Decision 345—Plant Breeders Rights, system based on UPOV, and (ii) Decision 391—A Common Regime on Access for the Andean Pact. This is the first law on ABS to adopted by a regional group. This regional framework provides inadequate protection of indigenous rights over lands and resources, but although it appears to go further than the CBD in recognizing the indigenous rights to control access to and use of their knowledge, this protection is not effective.[22] In the case of the ICBG in Peru, the indigenous populations were indeed consulted, but because of insufficient national regulation and institutional capacity in involving indigenous groups, the negotiations were highly problematic and it was mainly thanks to the assistance of NGOs and the good will of the ICBG actors that an end was achieved.

According to the Peruvian Constitution of 1993 'Renewable and non-renewable natural resources are the patrimony of the Nation'.[23] The attribute of natural resources as national patrimony suggests that they are not susceptible to alienation through the grant of exclusive rights. Genetic resources of indigenous lands are part of that national patrimony and the state may grant rights for their collection, subject to compliance with national and international law relating to the rights of indigenous peoples.[24]

There have been two sets of negotiations. The first one failed, but the second, with the lessons learned from the first one, succeeded in accommodating the positions of the parties. In the first set, the main agreements were the Letter of Intent (LoI) and the License Option Agreement. The second set includes the Biological Collecting Agreement (BCA), the License Option Agreement, and finally a know-how license between the local communities and the commercial partner Searle, a novelty in the bioprospecting practices.

[22] See Chapter 5 Section III.

[23] See also Decision 391 Article 6 and Chapter 5 Section III.

[24] Article 55 of the 1993 Constitution makes international treaties ratified by the State part of the national law. Both the CBD and the ILO 169 are hence part of the national legislation.

*Table 5: Agreement structure of the first round of the ICBG Peru negotiations*[25]

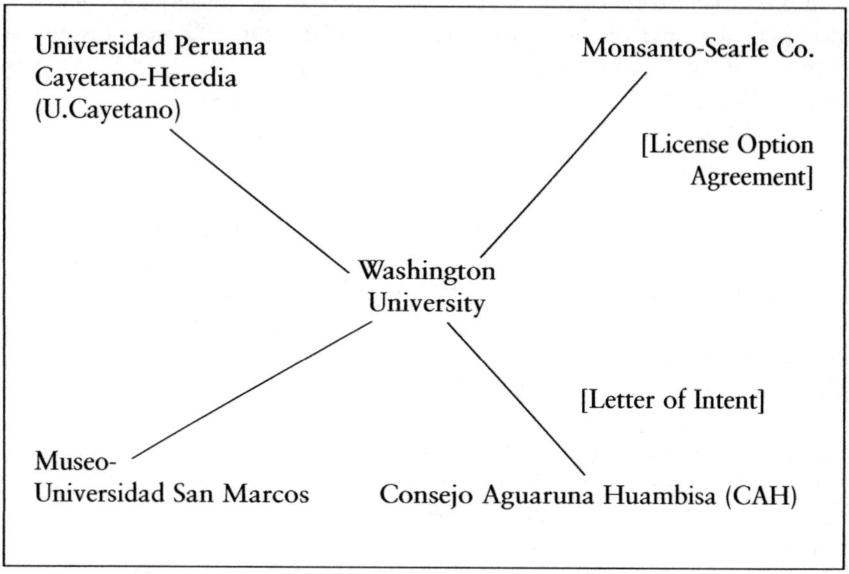


In the first arrangement, the LoI (letter of Intent) between CAH and WU set a precedent being the first time an indigenous peoples' representative organization was directly involved in negotiations relating to the ICBG Program. The LoI created the duty to 'protect and guarantee' ('*respetar y velar*') the genetic patrimony and/or the intellectual property rights of the local communities, and to take reasonable care not to injure these rights. The actual value, however, of such provisions is reduced by the fact that indigenous knowledge cannot generally be protected under existing intellectual property law. The responsibility for its protection falls upon the other party to the agreement, in this case the WU.

The objective of the License Option Agreement between WU and Searle was the collaboration in: 'identifying, inventorying, conserving and obtaining samples of ethnomedicinal plants and plant extracts from Peru, which are taxonomically diverse and/or have locally used medicinal applications, for the purpose of discovering new medicinally active compounds'. Among the terms of this collaboration, Searle would be entitled to exercise a right of first option over any developments arising from research and develop-

[25] B Tobin, presentation on the Peru ICBG project at the Workshop 'Rights to Plant Genetic Resources and Traditional Knowledge: Basic Issues and Perspectives' World Trade Institute, Bern (24–26 November 2002), on file with the author.

*Table 6: Agreement structure of the second round of the ICBG Peru negotiations*[26]

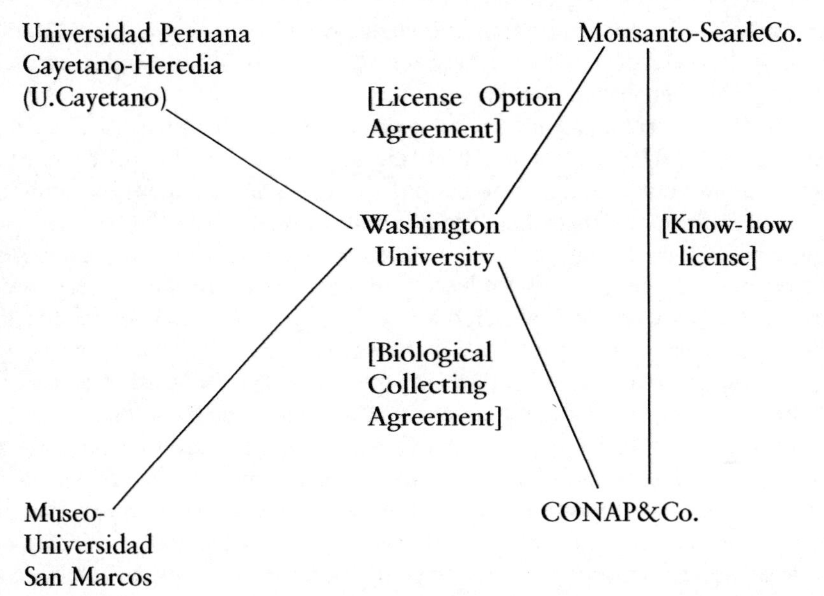


1. Confederación de Amazónicas Del Perú
2. Organización Central Communidades Aguarunas del Marañon (OCCAAM)
3. Federación Aguaruna Río Domingusa
4. Federación de Nativas Aguarunas del Río (FECONARIN)
5. Organización Aguaruna Alto Mayo (OAAM.)

ment work carried out by the universities. With regards to intellectual property rights, there was an implicit recognition that all indigenous know-how relating to plant extracts, or improvements is the intellectual property of indigenous peoples.

The first arrangements failed not only because of the questionable benefit-sharing settlement and the very low royalty rates, but also because of the serious doubts about the legitimate representation of the Aguaruna peoples by the CAH. A workshop took place in Lima, September 1995, to secure

[26] ibid.

wider discussion of the project. A decisive new actor was the NGO Sociedad Peruana De Derecho Ambiental (SPDA) providing valuable advice on the sharing of benefits amongst all communities and on designing a transparent negotiating strategy for the indigenous peoples. Finally, the responsibility for the traditional knowledge used by Searle was placed upon WU, and no direct agreement between the traditional knowledge owners and the commercial partner existed.

The second arrangement came to correct these shortcomings (see Table 6 on page 159). The BCA covers both the collection of biological resources, which should not endanger populations of medicinal plants, and the collection of indigenous know-how. Indigenous know-how is defined as 'the knowledge, innovations, practices, expertise and secrets of the Aguaruna people with regard to the use of biological resources for medicinal purposes'. Prior Informed Consent is required for both collections, to be provided from the Aguarunas; in fact, individuals must sign a PIC form prior to being asked to provide information, and community members may not be asked for information until the head of community has also signed a consent form.

Also, universities had agreed on a Code of Conduct, and they undertook to carry out all their activities in a manner consistent with the highest standards included in the guidelines and codes of ethics of the International Society of Ethnopharmacology (ISE), relating to collection of biological materials and traditional knowledge. The monetary benefits include royalties, up-front payments, license fees and milestone payments. The second arrangement simplified the way royalties should be calculated and defined how up-front payments were to be used.[27]

The Collaborating Organizations suggested a system of milestone payments within the know-how license to reflect the incremental value apportioned to traditional knowledge resulting from decisions to approve increased funding for research during product development, an licensing practice used in cutting edge industries, such as biotechnology and computer software, which provides the licensee with exclusive or non-exclusive rights to use informal knowledge that is not generally patentable, but it is important in the execution or utilization of an associated technology.[28] Such licensing agreements are frequently used to provide users with

[27] The initial License Option Agreement proposed a complex mechanism, under which royalty rates were determined depending on the extent to which the final product was developed using indigenous knowledge and the proximity between the use of the final product and the traditional use by indigenous peoples of the relevant biological material; the final project adopted a more simplified formula, and royalty levels were determined on the basis of whether the screening process for identifying useful active compounds in the development of a product is mechanistic (tests with no regard to traditional use, carried out by Searle) or functional (tests for activity similar to traditional use, carried out by the participating universities), see Tobin (n 21).

[28] World Intellectual Property Organization (WIPO) (ed) *Introduction to Intellectual Property, Theory and Practice* (WIPO, Geneva, 1997) 63–66.

a limited property interest, that is to say users make use of the relevant product or resource but they do not obtain a right of full ownership and must therefore terminate using the resource upon the termination of the license for its use, because the licensor retains an overarching proprietary right in the subject matter of the license.[29]

Traditional medicinal knowledge was seen as a key factor to identify resources for further research, and was treated as a form of information technology. Licensors (the collaborating local communities) place a number of obligations upon the licensees (the Searle company), such as free licenses to the licensor over any new technology developed (grant-back license). The BCA also foresees that:

> ... any application for patent or other intellectual property protection for a product developed for the same or similar purpose as the historic use by the Aguaruna people, or any part thereof, shall be the subject of a patent application which clearly recognizes the role of the Aguaruna People in the invention of the subject-matter of the intellectual property application.

The BCA sets out the mechanisms for the fair and equitable distribution of benefits among the collaborating organizations and the Aguaruna communities not parties to the agreement. A committee is formed to manage the funds which should be used for the financing of projects designed to promote education, health, conservation, sustainable management of natural resources, and other local and regional projects. The provision was novel because the benefit distribution mechanism was designed to ensure the widest possible participation of the Aguaruna communities.

Non-monetary benefits included scientific collaboration training, indigenous consultants, database software, access to pharmaceuticals in good faith, and reporting on R&D. In the end, despite this arrangement, no anti-malarial pharmaceuticals were developed that would activate the terms of the agreement, namely that 75 per cent of royalties, licence fees and other payments deriving from the patent or other income from research would return to Peru to support two universities and the well-being of the Aguaruna federations.

This case aptly shows that efforts to constructively involve indigenous communities can gravely complicate the access and benefit-sharing process, especially due to the lack of domestic regulation. The particular administrative and organizational schemes of indigenous communities need to be studied, every time there is a bioprospecting undertaking involving indigenous lands and knowledge.

[29] ibid.

For cultural and linguistic reasons, it is also difficult to communicate to indigenous peoples the whole bioprospecting process and explain that the resources and knowledge they provide will not always result in lucrative products. In the Peru ICBG the enthusiasm and very high expectations on monetary benefits on the part of the indigenous communities confused the negotiations and Searle terminated the know-how license. Further, indigenous and local languages might not have words to accommodate 'genes' and 'patents' jargon.[30] In fact, an interesting device in the ICBG Maya was the use of theater: theater plays an important role in the Maya culture. Locals were recruited to act in approximately 50 plays demonstrating plant collection and testing and how locals could contribute. Unfortunately, the project was heavily criticized and the companies like Diversa turned their interest to other countries, considerably reducing the prospects for bioprospecting ventures in this area in the near future.[31]

The lesson is that in today's 'social capital' economy, it is essential to invest in strong, long-term relationships based on mutual trust to foster development. The transfer of intellectual and intangible assets is contingent upon relationships: it is human networks that increasingly count today, and it is these networks that make things happen, not merely the inert, underlying data and information.[32] In this regard, anthropologists' expertise could be of great help in access and benefit-sharing ventures with indigenous peoples.

At the same time, increasing the focus on local diseases will make bioprospecting undertakings more meaningful and acceptable for locals. It is a shame that R&D for developing countries' ailments, like tuberculosis, malaria and other parasitic diseases, do not invite much investment compared to drugs and cosmetic devices in the developed world, because, unfortunately, decisions in drug companies depend on anticipated sales. In this respect, the ICBG Maya was indeed a missed opportunity, because it focused on locally important health needs promoting research on anti-malarial drugs. Since it is difficult to direct the market forces, focus on non-monetary benefits should be a priority: through capacity building local experts are taking the responsibility for the continuation of the project and the odds to identify medicines from natural products are improved. The bioprospecting undertaking can meet development goals, when the objective is to leave behind, after say 10 years of work, an organization that is fully prepared to find a drug.

[30] See Greene (n 21).

[31] NGOs criticized this effort claiming that the plays were used as smokescreen to obscure the researchers' real objective, commercial exploitation; 'This is a foregone opportunity for the Maya', said Ignacio Chapela, a plant molecular biologist at the University of California, Berkeley, see Greene (n 21); J Rosenthal, 'Curtain has Fallen on Hopes of Legal Bioprospecting—Local Communities, Too, Could Have Benefited from Better Health Care and Conservation' (2002) 416 Nature 15; and R Dalton, 'The Curtain falls' (2001) 414 Nature 685.

[32] A Krattiger, 'Financing the Bioindustry and Facilitating Biotechnology Transfer' (2004) IP Strategy Today 8–1.

Another interesting feature of the agreement relates to indigenous land rights. Under Andean and Peruvian law at the time of negotiation, indigenous and local communities had no ownership rights over genetic resources in plants or plant extracts. By making use of the biological material conditional upon the continuance in force of the know-how license, the BCA and the know-how license together extended the rights of indigenous and locals, represented by the Collaborating Organizations, to cover associated genetic resources.

## III. CORPORATE POLICIES

Parallel to institutional policies, corporate policies on access and benefit-sharing are significant unilateral instruments based on the good will of the 'strong' negotiating partner in the agreement. In their partnerships with public sectors institutions, private companies are obliged to comply with the public institutions' policies, such as the NCI. At present, the concepts of corporate social responsibility and liability are booming in scholar literature by 'looking at the activities of the economic enterprises beyond the horizon of strict regulation, taking into account the broader social context in which the enterprise operates and developing a bottom-up approach to the identification of the ethical and legal standards that a good and responsible corporation should follow in the pursuit of its economic mission'.[33]

From an international law perspective, taking into account that in today's globalized world multinational corporations are main actors at the international arena, their activities are guided by the shareholders' interests, and by the obligations deriving from their action in society as well.

From a law and economics perspective, there is a strong incentive for companies to avoid a large market reaction to the revelation of negative information with regards to their performance and transactions and their fear induces a higher level of compliance.[34]

[33] C Avery, Business and Human Rights in a Time of Change, in T Kamminga and S Zia-Zarifi (eds) *Liability of Multinational Corporations under International Law* (Kluwer Law International, The Hague, Boston, 2000) 17; see also R Mullerat (ed) *Corporate Social Responsibility: The Corporate Governance of the 21st Century* (Kluwer Law International, The Hague, 2005); F Francioni, 'Four Ways of Enforcing International Responsibility for Human Rights Violations of Multinational Corporations' paper presented at the Workshop 'Biotechnology, International Law and Human Rights, (unpublished manuscript, on file with the author).

[34] JS Johnston, 'Signaling Social Responsibility: On the Law and Economics of Market Incentives for Corporate Environmental Performance' (2005) U of Penn, Inst for Law & Econ Research Paper 05–16; see also the demonstration of the conflict between sustainable development and market liberalism through the Coca-Cola groundwater use in India by DA Kysar, 'Sustainable Development and Private Global Governance' 83 (2005) Texas Law Review 2109.

Corporate policies can have several levels, depending on the entity that defines and verifies the compliance criteria. The first level is corporate policies set by individual companies (first party), which set their own product and services standards. The second level comprises standards set by industry associations (second party), certification procedures within the industry or by external certifying companies. Finally, the third-level scheme is established by a private or public initiator, which is independent from the producers, distributors, and sellers of the products and services. All of these patterns can have third-party verification. Setting values and standards in their activities, multinational companies show willingness to comply with social and ethical rules, a very welcome development in corporate behaviour.[35]

The objective of a Code of Conduct (CoC)[36] from a specific company or industry is to signal information about their social and environmental behavior. To be effective, this information should be accessible and easy to understand for consumers, environmental and community organizations, and regulators. In any case, the information quality is enhanced when these voluntary schemes' requirements are established by independent bodies (third level) and the credibility and transparency of this information is better ascertained when verified by third parties.[37]

In access and benefit-sharing in the pharmaceutical sector, the Merck/INBio agreements are perhaps the most publicized.[38] In the botanical industry, Shaman Pharmaceuticals (currently NAPO Pharmaceuticals), made use of ethnobotanical knowledge for the identification of potential drugs. The Healing Forest Conservancy was the entity responsible for returning benefits to the communities. The initiative started with a lot of excitement, but in 1994 Eli Lilly, Shaman's pharmaceutical partner, did not renew the research contract. Andes Pharmaceutical was established in 1993 with the specific objective to assist source countries to add value in their

[35] S Wood, 'Voluntary Environmental Codes and Sustainability' in BJ Richardson and D Wood (eds) *Environmental Law for Sustainability* (Oxford and Portland, Oregon, Hart Publishing, 2006) 236; C Fischer et al, Corporate Codes of Conduct: Is Common Environmental Content Feasible? Resources for the Future, Discussion Paper 05-09 Conducted for the Foreign Investment Advisory Service of the World Bank Group, March 2005.

[36] From a legal terminology angle, the term 'Code of Conduct' connotes not only written rules set by private entities, as it is employed in the current thesis, but also rules on transnational business relations established by states and international organizations; in state practice, the use of the term is ambiguous and it is not easy to distinguish when a 'Code' refers to a recognized formal law-creating process or to a legally non-binding 'voluntary code'; see EU Petersmann, 'Codes of Conduct' in R Bernhard (ed) *Encyclopedia of Public International Law* (North Holland Amsterdam/New York, 2000) Vol I, 627.

[37] Fischer et al (n 35).

[38] See also the Merck Code of Conduct, *Our Values and Standards Edition II*, effective 7 November 2005.

resources.[39] Their business model involved biodiversity prospecting and drug discovery activities through the establishment of joint ventures in countries that are rich in biodiversity. The above stories of bilateral access and benefit-sharing agreements in the absence of a binding international and/or domestic legislation, demonstrate that under circumstances there is a degree of sensitivity on the part of private sector in social, environmental and here in biodiversity matters.[40]

At the international level, the International Labour Organization has issued clearly articulated standards in social matters, fundamental principles, and rights at work which have gained international acceptance. They are international treaties that are subject to ratification by the ILO members and contain recommendations of a non-binding character.[41] In environmental matters, however, it is not easy to agree on standards, because environmental standards depend on a greater diversity of variables, such as geography, climate, and the country's state of development. Developing countries usually prioritize poverty, basic health care, and education over the ecological and health risks handled by environmental regulations. Thus, environmental preferences vary and cannot indicate international convergence.[42]

The benefits of codes and standards rest within their ability to provide a reference framework. There is a broad range of existing corporate codes of conduct, International Organization for Standardization (ISO) family of standards[43] and ecolabels.[44] In environmental management, ISO 14001 certification series has been very successful.[45]

[39] Y-R Chen, 'Failed Attempt at Bioprospecting—A Case Study of Andes Pharmaceuticals and Implications' Sophomore College 29 September 2004.

[40] A benefit-sharing example in the motor industry is the following: in the early 1990s, Daimler–Benz set up the Poverty and Environment in Amazonia (POEMA) Programme in cooperation with the Federal University of Pará (UFPA) and the cooperative PRONAMAZON. A range of products made out of 'lexifrom' from natural fibres (oils and resins) and coconut fibre headrests and manufactured by PRONAMAZON in local communities around Belém are now fitted to C-class Mercedes-Benz, while the company envisages the expansion of the use of these materials to lorries, trains and aircraft manufacture and the transition from dependence to fossil resources to renewable resources. Working with local communities POEMA set activities in basic sanitation, health and education, multistrata agro-forestry, renewable energy and the processing of non-timber forest products. In 1999, DaimlerChrysler (the outcome of the merger of Daimler-Benz with Chrysler) announced a new partnership with the World Bank to expand public-private partnership in sustainable development activities. See RRB Leakey, Development of New Materials and Benefit-Sharing in the Motor Industry, in SA Laird (ed) *Biodiversity and Traditional Knowledge—Equitable Partnerships in Practice, People and Plants Conservation Series* (Earthscan Publications, London, 2002) 275 Box 8.10.

[41] See ILO, International Labour Standards, published on the Internet at: http://www.ilo.org/public/english/standards/norm/index.htm.

[42] Fischer et al (n 35) 2.

[43] ISO (www.iso.org), the world's largest developer of standards, is an international non-governmental framework of the national standards institutes of 150 countries; its central secretariat is in Geneva, Switzerland. The ISO was created in 1947 after a meeting in London, UK, where delegates from 25 countries came together to facilitate the international coordination and unification of industrial standards.

At the international level, there have been attempts to promote good corporate behaviour in social and environmental issues. Corporate social responsibility and accountability were addressed at the 1972 UN Conference on the Human Development, the 1992 UNCED in Rio and the 2002 WSSD in Johannesburg.[46] After the unsuccessful 1990 UN Draft Code of Conduct for Transnational Corporations, and following the address of United Nation Secretary-General Kofi Annan to the business leaders at the 1999 World Economic Forum to join an international initiative that would bring companies, UN agencies, and labor and civil society together to support universal environmental and social principles; the UN Global Compact: Advancing Corporate Citizenship was made operational at UN Headquarters on July 26, 2000. The Global Compact is far from a regulatory instrument monitoring the behavior of the companies and enforcing compliance; it is rather a voluntary initiative to advance responsible globalization, fight corruption, and promote human rights. In addition, in August 2003, the Sub-commission on the Promotion and Protection of Human Rights of the United Nations Commission on Human Rights adopted 'Norms on the Responsibilities of Transnational Corporations and Other Business Enterprises with Regards to Human Rights'.[47] These norms recognize that states have the primary responsibility to ensure the promotion and protection of human rights, but transnational corporations are also responsible for this promotion and protection.

[44] There is a wide range of national ecolabels today. The adoption of ecolabels started in the early 1990s, when environmental institutions, agencies and ministries of the environment promoted several ecolabel schemes with the goal to increase public awareness of products that made less impact on the environment. At present, governmental involvement has decreased, and the criteria development and criteria management of ecolabels is done by third-party independent organizations. The Global Ecolabelling Network (GEN) was created in 1994 as an association of third-party organizations that label environmental performance (www.gen.gr.jp). It sets objectives and basic ecolabel criteria to be met by its member organizations. Also, ISO has addressed ecolabelling in the 14020 series (environmental performance labeling).

[45] Within the EC, the Environmental Management and Audit Scheme (EMAS) is a reputed Environmental Management Systems Certification Programs (Council Regulation EEC No 1836/93 of 29 June 1993, restricted initially to industry sectors, and Regulation EC No 761/2001 of the European Parliament and of the Council of 19 March 2001, that opened the system to all economic sectors).

[46] E Morgera, 'From Stockholm to Johannesburg: From Corporate Responsibility to Corporate Accountability for the Global Protection of the Environment' (2004) 13 Review of EC & Intl Env L 2.

[47] Economic, Social and Cultural Rights, Norms on the Responsibilities of Transnational Corporations and Other Business Enterprises With Regard to Human Rights, Commission on Human Rights, Sub-Commission on the Promotion and Protection of Human Rights, UN Doc E/CN.4/sub.2/2003/12/Rev.2 (Geneva 26 August 2003); see also the accompanying Commentary, Economic, Social and Cultural Rights, Commentary on the Norms on the Responsibilities of Transnational Corporations and Other Business Enterprises With Regard to Human Rights, Commission on Human Rights, Sub-Commission on the Promotion and Protection of Human RightsUN Doc.E/CN.4/Sub.2/2003/38/Rev.2.

On the EC front, the European Commission has supported in the past the development of the Micro-organisms Sustainable Use and Access Regulation International Code of Conduct (MOSAICC) by the Belgian Coordinated Collections of Micro-Organisms (BCCM),[48] together with 16 other organizations from around the world.[49] This project started in September 1997 for a duration of 21 months. It is a voluntary Code of Conduct, a tool to support the implementation of the CBD at the level of microbes and other micro-oganisms, in accordance with other relevant international and national rules. At present the Commission is financing a follow-up to the MOSAICC project aimed at providing validated reliable methods for the value assessment of microbial resources. Such methods are necessary to put a socially, economically, and environmentally 'sound' price on genetic resources and therefore facilitate benefit-sharing. The project also aims to develop validated model documents to enable traceability of microbial resources (origin, transfer and transport).[50]

Also, the EMAS certification scheme has a potential for private enterprises complying with the Bonn Guidelines.[51] EMAS '...is a voluntary scheme for organizations willing to commit themselves to evaluate and improve their environmental performance'. The EC stated that 'such a scheme would serve the purpose of helping users to improve their overall environmental performance, including in relation to access and benefit-sharing, but would not alter their legal obligations'.[52]

Considering the growing environmental governance with a great number of MEAs signed during the last decades and the pivotal role of the private sector in activities that impact upon environmental performance, the voluntary corporate behaviour is a rather promising development. A 2006 UN Report shows that even if there is still much to be done, there are a growing number of business and industry groups making efforts to improve their environmental and social performance.[53] With regard to access and benefit-sharing, good private practice has so far inspired national legislation.

[48] Micro-Organisms Sustainable use and Access regulation International Code of Conduct —MOSAICC (published on the Internet at: http://www.belspo.be/bccm/mosaicc).

[49] Among others OECD, IUCN, Royal Botanic Gardens KEW, INBio.

[50] See CBD Note by the Executive Secretary, *Secretariat analysis of measures to ensure compliance with prior informed consent of the contracting party providing genetic resources and mutually agreed terms on which access was granted, and of other approaches, including an international certificate of origin/source/legal provenance*, Doc. UNEP/CBD/WG-ABS/3/5, 10 December 2004.

[51] Communication from the EC to the European Parliament and the Council, 'The Implementation by the European Community of the 'Bonn Guidelines' on Access to Genetic Resources and Benefit-sharing under the Convention on Biological Diversity', Brussels, 23.12.2003, COM(2003) 821 final, 22.

[52] ibid.

[53] Industries trying to improve UN News Center, Environmental, Social Performance—UN Report, New York, May 10 2006; United Nations Environment Programme, *Class of 2006: Industry Report Cards on Environment and Social Responsibility* (UNEP, Nairobi, 2006).

However, corporate practices, lacking 'teeth', cannot replace domestic legislation, but only complement it.

## IV. PROFESSIONAL SOCIETY STANDARDS

A number of professional and research societies in fields like medicine, psychology, sociology,[54] archaeology[55] and social work are quite sympathetic to ethical issues and developed documents (statements of ethics, codes of ethics, codes of practice, guidelines and research protocols) to articulate ethical values standards for best practice in research activities. These ethical principles and guidelines evolve within the research/professional circles in parallel to the broader social, ethical and legal changes; researchers and professionals abide by the existing ethical considerations and respond to the societal reality by developing an 'ethical culture' in their profession.[56]

Professional standards comprise codes of ethics in conjunction with research guidelines. A code of ethics is a public moral system that encourages, requires, or prohibits certain forms of behavior according to rational and ethical criteria; it thus articulates underlying ethical values and the philosophical basis for research and practice. In turn, research guidelines detail current standards of best practice in research and are most effectively written to deal with research in a special context; they complement codes of ethics by providing practical details for the implementation of these values. Enforcement of the codes and guidelines by the professional groups and researchers varies. Some constitute mere recommendations upon discretion of the researcher and some other have a binding character on all the members of the professional society and are punishable (eg removal of rights to practice a profession).

[54] See eg American Sociological Association at http://www.asanet.org/cs/root/leftnav/ethics/code_of_ethics_table_of_contents (Code of Ethics).

[55] See eg Register of Professional Archaeologists http://www.rpanet.org/displaycommon.cfm?an=1&subarticlenbr=3 (Code of Conduct and Standards of Research Performance).

[56] Ethical codes are adopted at the level of federal-funding bodies, such as the 'Tri-Council Policy Statement on Ethical Conduct for Research Involving Humans' that governs all research supported by the Medical Research Council of Canada (MRC), the Natural Sciences and Engineering Research Council of Canada (NSERC) and the Social Sciences and Humanities Research Council in Canada (SSHRC). The Statement includes sections on free and informed consent, privacy and confidentiality, and research involving aboriginal peoples. There is a top-down and a bottom-up approach in the Statement: on the one hand, investigators who receive federal research funding in Canada are compelled to be aware of the Tri-Council policy statement and insure that their research complies with its demands; on the other hand, researchers are given the implicit responsibility and opportunity to influence ethical policies at the federal and institutional levels through a 'bottom-up' approach, see SA Laird and DA Posey, 'Professional society standards for biodiversity research: codes of ethics and research guidelines' in Laird (ed) 2002 (n 40) 31 Box 2.6 by Kelly Bannister.

In the area of access and benefit-sharing, a number of codes and/or guidelines have been developed in the related fields of anthropology, ethnobiology, pharmacognosy and ecology,[57] which prioritize certain principles, rights and obligations in the conduct of research and in professional practice, such as the respect of indigenous communities' rights, transparency, honesty and confidentiality. A list of codes include the International Society of Ethnobiology (ISE) 1998 Code of Ethics and Guidelines for Research, Collections, Databases and Publications; the American Anthropological Association (AAA) 1998 Code of Ethics;[58] the American Society of Pharmacognosy (ASP) 1992 Guidelines for Members on issues associated with natural products; the Society for Economic Botany (SEB) 1995 Guidelines of Professional Ethics of the Society for Economic Botany; the 1996 Pew Conservation Fellows Biodiversity Research Protocols: Guidelines for Researchers and Local Communities interested in Accessing, Exploring and Studying Biodiversity; and the 1992 Manila Declaration concerning the Ethical Utilization of Asian Biological Resources.[59]

In fact, the ISE is the first international federation representing the full interdisciplinary field of ethnobiology. It was established in Belém (Brazil) in 1988 on the occasion of the First International Congress on Ethnobiology, when indigenous and local communities met with scientists and environmentalists to discuss a common strategy to stop the rapid decrease in the planet's biological and cultural diversity. The outcome of the Congress was a statement of principles for ethnobiology and ethnobiologists in the international context, widely known as the Declaration of Belém. The Declaration is an awareness-raising tool incorporating in a broad language calls for prior informed consent, benefit-sharing and collaborative and equitable relationships with the indigenous peoples. In

[57] CBD Doc UNEP/CBD/WG-ABS/3/5; Laird and Posey in Laird 2002 (n 40) 16 f.

[58] See http://www.aaanet.org/committees/ethics/ethics.htm (23/12/08). In 1965 there was strong public outcry over US government support of social science research in foreign countries; the Project Camelot in Chile, where the army was believed to fund social science research to assess potential for internal war and prevent public revolts against the Chilean government, or the war in Vietnam, raised ethical questions and damaged the reputation of anthropological sciences. Consequently, the AAA formed a Committee on Research Problems to draft a Code of Ethics 'Statement of Problems in Anthropological Research and Ethics', adopted in 1967; later a Standing Committee on Ethics was established and in 1970 a new draft Code of Ethics was presented.

[59] See the detailed table Table 2.1 with the elements of codes of ethics and research guidelines, by Laird and Posey in Laird (ed) (2002) 33–49; professional Ethics Resources on the World Wide Web are available at http://www.ethicsweb.ca/resources/ (23/12/08); also, in 1997 Tegan Churcher for the Biodiversity and Ethics Working Group on Pew Conservation Fellows compiled a Directory of Guidance Documents Relating to Biodiversity and Cultural Traditional Knowledge Research and Prospecting, Department of Geography, University of California and Environmental Energy Technologies Division, published on the Internet at: http://www-geography.berkeley.edu/ProjectsResources/BRP/SurveyA-F.html (23/12/08).

1998, a refined code of ethics, delineating principles to be followed in ethnobotanical research,[60] and research guidelines for action specifying 'best practice' issues related to data management, cultural and genetic resources collections and publications[61] was developed.

In response to concerns raised in Asia against the behaviour of foreign scientists collecting biological materials for pharmaceutical companies, the Manila Declaration was adopted at the Seventh Asian Symposium on Medicinal Plants, Spices and Other Natural Products (ASOMPS VII), 2-7 February 1992, Manila (Philippines). The Declaration sets a series of recommendations based on the principle of collaboration and includes two Appendices, a 'Code of Ethics for Foreign Collectors of Biological Samples' and 'Contract Guidelines' for the provision of biological materials. The four basic aspects of the Declaration are: the natural biological heritage of each country should be respected and explored for the benefit of that country (royalty sharing including); collaboration with foreign scientists should be conducted on a mutually acceptable basis and, where possible, should include training of local scientists and technology transfer; legislation should be developed to cover the collection and export of materials, and equitable royalty and license arrangements; and traditional knowledge must be recognized as significant intellectual property.[62]

Despite the passionate endorsement of the Declaration from the Asian scientific community, its impact was not considerable, and the Melaka Accord was adopted in 1994 by ASOMPS VIII (12–16 June 1994, Melaka (Malaysia). The Accord emphasizes the need for a binding instrument, national legislation or even regional regulatory instrument among the ASEAN countries. The Accord asks also from the editors of scientific journals to require that publications based on material from outside their country to cite in the acknowledgements the permit approvals, a responsibility widely ignored by editors.[63]

[60] International Society on Ethnobiology—Code of Ethics, published on the Internet at: http://ise.arts.ubc.ca/global_coalition/ethics.php (23/12/08) The Code refers to fifteen fundamental principles: Principle of Prior Rights, Principle of Self-Determination, Principle of Inalienability, Principle of Traditional Guardianship, Principle of Active Participation, Principle of Full Disclosure, Principle of Prior Informed Consent and Veto, Principle of Confidentiality, Principle of Respect, Principle of Active Protection, Principle of Precaution, and the Principle of Compensation and Equitable Sharing.

[61] International Society on Ethnobiology—Guidelines for Research, Collections, Databases, and Publications, published on the Internet at: http://guallart.dac.uga.edu/ISE/iseResearch.html (12/05/2005).

[62] See Laird and Posey in Laird (n 40) Box 2.5; The Manila Declaration and Melaka Accord by Ruth Kiew *(Botany 2000 Asia)* 29; PR Andrews et al., 'Preservation and Utilization of Natural Biodiversity in Context of Search for Economically Valuable Medicinal Biota' (1996) 68 Pure Applied Chemistry 2325–2332.

[63] Laird and Posey in Laird (n 40) 30, Box 2.5 by Ruth Kiew; Andrews et al, (1996).

Following the call of the Melaka Accord for the development of a common form of legislation, an informal meeting of ASEAN scientists took place in Kuala Lumpur. Today, only the Philippines[64] and the state of Sarawak, Malaysia, have come up with legislation, but the lack of a regional framework can be a disincentive for investments.

## V. CONCLUSION

Access and benefit-sharing standards developed within the private sector fill the gaps within national/regional and international rules. Indeed, private practice, outlined above in institutional policies, professional societies' standards, and more importantly corporate policies, can prove to be a flexible and practical approach to implement the CBD objectives and help changing legal regimes and research relationships.

Actually, such practices form part of the user countries' obligations to implement the CBD; the legal basis is found in CBD Article 15.7 of the CBD, which foresees that '*each* Contracting Party shall take legislative, administrative or policy measures...with the aim of benefit-sharing' (*emphasis* added) and further specified by the Bonn Guidelines list responsibilities that user countries have with regard to access and benefit-sharing. To this end, a continuum of measures by the public (government) and private/professional sector are envisioned. In fact, public measures prove a rather difficult task and most of the burden for access and benefit-sharing inevitably falls on the providing country, due to the often multinational character of drug companies, but more administrative measures and other incentives for good behavior can be of great assistance. However, with the rising influence of civil society and the power of the media, consumers expect a more conscious and sensitive behavior from private companies. Professionals, companies and institutions, as well as international organizations engaged in economic and investment activities,[65] want to avoid 'environmental blacklisting', contributing to an evolving body of voluntary compliance.

The combination of codes of ethics and research guidelines has been suggested as the most effective approach to address the range of issues raised by biodiversity research.[66] The success, however, of this route is

[64] See Chapter 5.

[65] See The World Bank, *Focus on Sustainability 2004* (Washington DC, The World Bank, 2005); the report is the first of its kind completed by a multilateral development bank; although the World Bank is not a private corporation, its activities resemble the activities of the private sector and this report looks at the World Bank's activities through the lens of corporate social responsibility reporting criteria.

[66] Laird and Posey in Laird (ed) (n 40) 21.

questionable, because committees of ethics lack institutional capacity and 'teeth'. At the same time, an important lesson is that international and domestic policy, as it stands today, is inadequate as a sole mechanism to control access to genetic resources and determine variables for benefit-sharing. Domestic legislation in many cases has elaborated workable access and benefit-sharing prescriptions on the basis of what has emerged from the MATs, research, and commercial agreements.

One consistent conclusion from private practice, which predates domestic legislation and has been continuously refined through time, is that it is very difficult to generalize about bioprospecting; a 'one size fits all' international instrument to regulate access to genetic resources and benefit-sharing by all bioprospecting-active industries is a very difficult mission. States with user companies and institutions within their jurisdiction are expected to advance and facilitate such voluntary corporate behavior. In fact, the European Commission is a pioneer in this sector: a European network of access and benefit-sharing focal points and national authorities connected through the European Community Biodiversity Clearing-House Mechanism has been created,[67] where a specific section on ABS is envisaged to contain the text of the Bonn Guidelines accompanied by explanatory notes for the different industries, together with a register of stakeholders. Users may be informed about their international rights and obligations in relation to ABS not only in the framework of the CBD, but also of the FAO-IT. Also, the European Commission has created an inter-departmental working group on indigenous issues to promote EC action in this field and ensure its coherence throughout the wide array of policies relevant to indigenous peoples, including environment, development, cooperation, human rights, trade and intellectual property.[68] Consequently, concurrent policies developed in the private domain express the general ethos of individuals, institutions, and corporations involved in bioprospecting offer useful guidance in access and benefit-sharing policy and complement access and benefit-sharing regulation, but such private instruments cannot fully articulate the details of negotiated agreements. Their nature is complementary to the necessary access and benefit-sharing legislation, which is expected to effectively guide future foreign collaborators and strengthen the bargaining capacity of local institutions.

[67] See http://biodiversity-chm.eea.eu.int/ (14/02/02).

[68] CBD Doc UNEP/CBD/WG-ABS/3/5.

# CHAPTER 7

# *Policy Evaluation*

## I. INTRODUCTION

The degree of stakeholders' participation in the access and benefit-sharing undertaking determines whether its process and outcome are 'fair and equitable' following the call of the CBD. Among the stakeholders two main categories are discernible: direct and indirect.

The definition of indirect stakeholders might not seem very important at first glance. However, defining the indirect stakeholders helps us view the 'big' picture of bioprospecting, namely its significance for biodiversity conservation and wider health care, detached from narrow commercial interests and intellectual property concerns. Indirect stakeholders include all of humanity and people who suffer from a specific ailment. Indirect stakeholders are not entitled either to grant PIC or to have claims on the deriving benefits; they benefit from the constructive use of the medicinal plants that will end up with a successful drug and also from the conservation of a healthy ecosystem.

Contrary to the broad and general definition of the indirect stakeholders, the identification of the direct stakeholders is inherently related to the contractability definition of bioprospecting. The direct stakeholders we usually encounter in the bioprospecting contracts are government/State agencies, indigenous populations/local communities,[1] public citizens' stakes/NGOs, industry and research and science; they arise by the recognition that they are all holders of rights over the biodiversity goods and the biodiversity derived goods. This complex group of stakeholders derives from the contention here that neither a predominately private nor an exclusively public regulatory approach can prove effective in access and benefit-sharing. The avenue of regulating the entire undertaking through an exclusively contractual approach runs diverse risks against the interests of social groups such as indigenous peoples and cannot be a remedy for environmental quality enhancement, whereas a highly interventionist line of action by State

[1] It has been commented that indigenous peoples' characterization as 'stakeholders', together with corporations, academic institutions, NGOs, etc, is demoting and that indigenous peoples should have the status of 'rights' holders', see D Harry and LM Kanehe, 'The BS in Access and Benefit-Sharing (ABS): Critical Questions for Indigenous Peoples' in B Burrows (ed) *The Catch: Perspectives on Benefit-Sharing* (Edmonds Institute, Edmonds, Washington, 2005) 81.

authorities may ignore the already existing big markets of biodiversity goods. A successful access and benefit-sharing policy recognizes that market creation alone is an inadequate tool to achieve biodiversity sustainable use and conservation, but harnesses the existent market power and directs it towards these ends.

This hybrid biodiversity policy model between public and private law implies that the state is expected to foster biodiversity markets by establishing legal instruments that the private sector can value and exchange,[2] while at the same time a specialized state authority coordinates bioprospecting undertakings within the country, protects the weak parties to the contracts, and guarantees that the benefits are channelled to those responsible for the protection of the said resources. The establishment of a clear, enforceable property rights regime is fundamental for the definition of the direct stakeholders, in other words the parties which grant the PIC, participate in the contracts negotiation and claim benefits in the event of successful commercialization.

Since we recognize that clear property rights are a prerequisite for successful access and benefit-sharing policy, this chapter first explores the property-related aspects of bioprospecting, identified with relation to lands and resources, to genetic material, and finally to traditional knowledge (intellectual property). Then, it takes a look at the indigenous peoples' rights sketched in current international law. The analysis is necessary to highlight the obligation of the States to ensure an improved position of indigenous peoples, as stakeholders in the bioprospecting undertakings who are consulted and compensated, by recognizing their rights to land, resources and knowledge and also acknowledging their customary laws. States rich in biodiversity which can profit from bioprospecting opportunities happen to host large indigenous populations within their borders; indigenous societies have to date developed sophisticated systems of biodiversity conservation, use, and management; and failure to properly recognize their rights over the resources they manage and the knowledge they provide renders access and benefit-sharing policies unfair and inequitable. Indigenous populations have been considerably ignored in access and benefit-sharing practice so far, and it is here supported that access and benefit-sharing policies should follow the current developments in human rights law.

A short overview of the bioprospecting process and benefits distribution follows with a view to highlighting the difficulties arising from contract practice and to emphasize the importance of non-monetary benefits.

[2] See also Chapter 2; OECD, *Handbook of Market Creation for Biodiversity—Issues in Implementation* (OECD, Paris, 2004); OECD, *Harnessing Markets for Biodiversity—Towards Conservation and Sustainable Use* (Paris, OECD, 2003); OECD, *Handbook of Biodiversity Valuation—A Guide for Policy-Makers* (Paris, OECD, 2002); OECD, Handbook of Incentive Measures for Biodiversity—Design and Implementation (Paris, OECD, 1999).

Finally, international relations theory is employed confirming the empirical results on the immaturity of the international environment for a universal binding access and benefit-sharing treaty endorsed by developing countries and other organizations.

## II. BIOPROSPECTING AND PROPERTY RIGHTS CONSIDERATIONS

We saw that property rights constitute the main legal instruments for a working market in bioprospecting: lack of property rights or ill-defined property rights create high-level risks and uncertainty. Property rights express the exclusive authority to determine how and by whom a particular resource is used, and hence they facilitate contractual activity.

Property rights appear in different forms in different societies and legal systems and engage complex legal and institutional dimensions: they may be seen as a bundle of separate and distinct rights over a particular good/asset, including at least the right of personal use, the right to demand compensation as a prerequisite for its use by other people, and the right to transfer any or all of these rights to others (either permanently by sale or temporarily through some form of contractual arrangement).[3]

The CBD acts as a framework convention and calls for further State implementation, hence States are further required to determine the legal status of lands (property, land tenure and trusteeship issues), and of natural and genetic resources within their territory. State decisions on property rights are decisive for the capacity of the domestic access and benefit-sharing regime to meet the CBD objectives.[4]

Property rights do not necessarily imply that a good or service is privately owned; governments may exercise property rights through their designated officials (public property/ownership) and private individuals and other sorts of non-governmental organizations (private property) may also exercise property rights.[5]

Property rights should not merely be understood as conceived in the Western civil and common law systems. Some societies have developed informal, alternative property rights (communal ownership), which although not backed by formal laws, are very clear. For instance, indigenous peoples' communities have very well defined property rights institutions in their customary laws. The major challenge here for the domestic policy makers is to recognize these customary laws that differ from the western-style property institutions, understand them, and make use of

[3] OECD 2004 (n 2) 27.

[4] W Lesser, *Genetic Resources—Convention on Biological Diversity: Exploring Access and Benefit-Sharing Issues* (CAB International, Oxon, New York, 1998),13 f.

[5] OECD 2004 (n 2).

them in the bioprospecting contracts. In bioprospecting activities three different types of property rights are discerned where the state is called to define their legal status: property over lands and natural resources, over genetic material and, finally, intellectual property.

### *A. Property Rights to Lands and Natural Resources*

Genetic material is extracted from natural resources within lands belonging to an individual, to a community, or to the State, where these lands are public, such as national parks and public collections. The CBD confers sovereignty over resources and the countries have to decide property matters over natural resources and lands. Property rights of individuals and entities should be clearly defined at the domestic level to ease the bioprospecting negotiations and contracts. But this is not actually a very easy exercise.

International documents, including the CBD, emphasize the importance of property and land rights for the protection of the environment and development. For sustainable management in planning and managing land resources, Agenda 21 acknowledges the significance of 'private property rights, the rights of indigenous people and their communities and other local communities'.[6] Also, with a view to promoting land conservation and improved management and income diversification in Sustainable Agriculture and Rural Development, Agenda 21 encourages governments to 'assign clear titles, rights and responsibilities for land and for individuals or communities', to 'ensure equitable access of rural people, particularly women, small farmers, landless and indigenous people, to land, water and forest resources' and to 'implement policies to influence land tenure and property rights positively with due recognition of the minimum size of land-holding required to maintain production and check further fragmentation'.[7]

The CBD itself protects the 'customary use of biological resources in accordance with traditional cultural practices' (CBD Article 10(c)). While this provision has been considered to be strongly related to Article 8(j) on the protection of traditional knowledge, traditional uses and practices in

[6] Agenda 21 Chapter 10.

[7] Agenda 21 Chapter 14. The importance of land and property rights are further highlighted in Agenda 21 in Chapter 3 on combating poverty, Chapter 7 on human settlements and Chapter 12 on combating desertification. Other international efforts include the Rome Declaration on World Food Security and World Plan of Action, 13–17 November 1996, Rome, Italy, where the interrelationship between access to land, environmental sustainability and the fight against hunger was stressed, and the Eighth Session of the Commission on Sustainable Development in New York (April and May 2000) which highlighted the importance to land access and tenure issues. See also the detailed analysis of human rights instruments described at the Indigenous Peoples' Rights (Section 3 below).

the management of biological resources should be respected and encouraged.[8]

What happens is that many developing countries rich in biodiversity lack sufficient property rights institutions. Development institutions and academics insist on the elaboration of clear land law structures and land tenure systems as incentives to conserve and produce and, thus, as a qualification for development, foreign investment, and prosperity.[9] There is currently a land law reform trend worldwide sustained by development institutions and other organizations, for the substantive definition and administrative operation of property rights.[10]

However, property rights reform appears to be a very tricky issue. First, in many developing countries there are already customary property and land tenure institutions with ambits clearly defined; such institutions are usually developed by the large indigenous groups within their territories.[11] The characterization of such systems as 'backward' led to the implantation of statutory systems during the colonial era resulting in concomitant customary and statutory laws and a difficult to manage regulatory situation today. Statutory systems actually failed to replace existing customary rules because property institutions are grounded on cultural and social values.

Secondly, in a significant number of countries the State remains the ultimate titleholder, with private rights over land limited to usufructs, long-term leases or permits. Property rights reforms signify constitutional reforms and ideological clashes. Moreover, the simple recognition of ultimate state ownership is not adequate to evaluate the private rights quality, because the nature and strength of such State ownership varies from nominal and symbolic to active and interventionist.

Thirdly, the increased protection of indigenous peoples' rights[12] calls for the recognition of customary land tenure and property systems, but the overlapping of indigenous and customary laws within the country complicate the efforts for a safe legal and administrative environment. There are

[8] CBD Note by the Executive Secretary, *Traditional Knowledge and Biological Diversity*, Workshop on Traditional Knowledge and Biological Diversity, Madrid, Spain (24–28 November 1997), Doc UNEP/CBD/TKBD/1/2, 18 October 1997.

[9] H De Soto, *The Mystery of the Capital: Why Capitalism Triumphs in the West and Fails Everywhere Else* (Bantom Press, New York, 200); see the Oxfam web site on land rights in Africa at: http://www.oxfam.org.uk/resources/learning/landrights/index.html (23/12/08) and LandNet Americas on promotion of property rights registration in Latin American countries at: http://www.landnetamericas.org/index.asp (23/12/08).

[10] ibid. See also Instituto Libertad y Democracia (ILD), a private, non-profit organization headquartered in Lima, Peru founded by de Soto (http://www.ild.org.pe) and J. Lindsay, 'Chapter 8. Land' in FAO Legal Office *Law and Sustainable Development since Rio: Legal Trends in Agriculture and Natural Resource Management* FAO Legislative Study 73 (Rome: FAO, 2002) 203.

[11] R Roldán Ortiga, Models of Recognizing Indigenous Land Rights in Latin America, The World Bank Environment Department (The World Bank, Washington, DC, 2004).

[12] See Section 3.

several examples of domestic legislation responding to the international calls for further recognition of indigenous peoples' land rights and recognizing community-systems that better reflect the complex local social values and that provide tenure security. However, additional problems are present related to the definition of the elusive term of 'community':[13] recognition of community-based rights and their integration in the formal national land tenure system implies definition of the individuals who form the community, the exact land area over which legitimate tenure rights[14] exist and representation and decision-making processes according to the community institutions.[15]

Finally, ideological and political variables, such as social class conflicts and power over resources, political representation and corruption may further complicate the land reforms and indigenous movements' efforts. The need for wide public consultation and diplomatic means is needed[16] to boost land reforms and indigenous peoples' rights.

### B. *Property Rights to Plant Genetic Material*

The issue of property rights over the genetic material itself is actually rather intricate and controversial. The CBD does not decide on whether genetic resources are governmental property or not, but it gives authority to the governments to decide on the property regime of genetic resources within their territory.

[13] See Judgment of the Inter-American Court of Human Rights in the Case of the Mayagna (Sumo) Community of Awas Tingni v the Republic of Nicaragua, IACHR, Series C, No 79 (31 August 2001): 'Given the characteristics of the instant case, some specifications are required on the concept of property in indigenous communities. Among indigenous peoples there is a communitarian tradition regarding a communal form of collective property of the land, in the sense that ownership of the land is not centered on an individual but rather on the group and its community. Indigenous groups, by the fact of their very existence, have the right to live freely in their own territory; the close ties of indigenous people with the land must be recognized and understood as the fundamental basis of their cultures, their spiritual life, their integrity, and their economic survival. For indigenous communities, relations to the land are not merely a matter of possession and production but a material and spiritual element which they must fully enjoy, even to preserve their cultural legacy and transmit it to future generations.'

[14] Community based natural resources management (CBNRM), for example, is a people-centered approach in property management by indigenous communities and studies show that boundaries of indigenous territories alone are not the right indicator for the definition of exclusion and inclusion rights at the national level legislation, but that customary and other factors further determine the community members' definition and, thus, the establishment of whom to exclude from natural resource benefits should be left upon these customary norms; see AW Mosimane and KM Aribeb, 'Exclusion through Defined Membership in People-Centered Natural Resources Management: Who Defines?' Centre for Applied Social Sciences/Programme for land and Agrarian Studies (CASS/PLAAS) 2005.

[15] S Greene, 'Indigenous Peoples Incorporated? Culture as Politics, Culture as Property in Contemporary Bioprospecting Deals' (2004) 45 Current Anthropology 211.

[16] See 'The World Bank' *The Economist* (2 March 2006) and 'The World Bank and Corruption' *The Economist* (2 March 2006).

More specifically, following CBD Article 3:

> States have, in accordance with the Charter of the United Nations and the principles of international law, the sovereign right to exploit their own resources pursuant to their own environmental policies, and the responsibility to ensure that activities within their jurisdiction or control do not cause damage to the environment of other States or of areas beyond the limits of national jurisdiction.

Furthermore, CBD Article 15.1 reads: 'Recognizing the sovereign rights of States over their natural resources, the authority to determine access to genetic resources rests with the national governments and is subject to national legislation'. These CBD provisions explicitly recognize the sovereign rights of States over the resources within their jurisdictional control, and at the same time they confer the international obligation to the States to ensure that activities related to these resources do not harm the environment; moreover, the delineation of property and control over genetic resources in particular rests with the States, just like with exhaustible resources.

The attribution of sovereign rights over natural resources in general is not a novel idea: international practice has recognized sovereign control over exhaustible resources, like minerals and oil found within national borders, a practice broadly followed by states.[17] But genetic resources differ from other natural resources: genetic resources are 'contained' in natural resources, if one may say so, but most importantly genetic resources are of an intangible and informational character. The informational character of genetic material impedes its value quantification and consequently aggravates the complexity of policies to address bioprospecting compared to other biodiversity markets that have a clear use value.

Property rights to genetic material were not really discussed until the rise and progress of biotechnology and its application to a very large array of activities to cover human needs. The main question is who actually makes the decision for the treatment of a resource as 'commons' and whose interests are to be sought. History reveals that it is the economy that directs the transfer of wealth from public to private hands: the agricultural revolution and the importance of controlling land in the 18$^{th}$ century led to the enclosure of common lands; the industrial revolution in the 19$^{th}$ century piloted the expansion of intellectual property systems, and now, the information technology, biotechnology and nanotechnology are the cutting edge sectors

[17] Charter of Economic Rights and Duties of States, UN GA Resolution 3281 (XXIX) of 12 December 1974.

that will determine the property institutions that best fit the business practice.[18]

Obviously, property models exist and will continue to exist as long as there is a market and there is the need to value and exchange matter and information. However, there is a trend to allocate extensive property rights, leading to fragmentation and overlapping property rights and generating transaction costs and heterogeneous interests, a phenomenon very present in the world of medicines R&D and the overlapping intellectual property rights and extension of monopoly rights.[19]

In access and benefit-sharing contracts, proliferated property rights can eventually act as a disincentive for drug companies to invest[20] because in the end source countries tend to overestimate their biological assets. It was shown that in practice States opt for a State ownership system in genetic material either by law or by constitution: in their access and benefit-sharing laws, by reserving for the governmental authority a permanent and strong stakeholder status in bioprospecting activities, States assume property rights over the genetic material derived from natural resources within their territories; also, some constitutions expressly foresee that genetic resources are either patrimony or property of the nations.[21] Expansion of property rights over the biological matter and the information associated to the matter create a negative investment environment and bioprospecting fails to serve the public goals of sustainable use and conservation for which it is conceptualized.

To combat this property entropy,[22] biodiversity-rich States should put the public goals of biodiversity conservation and sustainable use and of biomedical research above privatization attempts, lessen their monetary expectations from bioprospecting, and try to promote the physical unity in both the cases of biological matter and subsoil resources, despite the different character of the two. Actually, the Roman tradition suggested that 'who ever owns the land owns the property all the way to heaven and all the way to the center of the earth',[23] but the 20$^{th}$ century codes abandoned this

[18] J Hepburn, 'Reclaiming Commons—Old and New' (2005) Presentation to the *Other World's Conference*, Sydney, April 2005, published on the Internet at: http://www.mercury.org.au/PDFs/Reclaiming%20Commons%20-%20John%20Hepburn.pdf (9/12/09); De Soto (n 9).

[19] M Heller and RS Eisenberg, 'Can Patents Deter Innovation? The Anticommons in Biomedical Research' (1998) 280 *Science* 696.

[20] S Safrin, 'Hyperownership in a Time of Biotechnological Promise: The International Conflict to Control the Building Blocks of Life' 98 (2004) American Journal of International Law 4, 641; see also M Heller 'The Tragedy of AntiCommons: Property in the Transition from Marx to Market' (1998) 111 Harvard Law Review 621.

[21] See Chapter 5.

[22] See F Parisi, 'The Fall and Rise of Functional Property' (2005) *George Mason Law & Economics Research Paper No. 05-38*, November 17, 2005, 18.

[23] ibid 18–19: 'Cuius est solum eius est usque ad coelum et usque ad inferos' Latin maxim.

roman rule of physical entity,[24] a movement criticized by mid-20th century civil law scholars for the violation of the modern ideal of unified property.[25]

With regards to subsoil resources in particular, the State ownership system has been criticized as 'distinctly discriminatory' in its operations in relation to indigenous peoples,[26] with the additional argument that indigenous peoples did not participate in the States' process for the adoption of the constitution and there is no data available that they consented to transfer their subsurface resources to the state.[27] Recent regional[28] and domestic case law[29] lends support to the inclusion of subsurface resources in the understanding of the natural resources scope over which indigenous peoples may assume property rights.

Alternatively to property rights, taxing the resources' users or setting zones and individual transferable quotas such as in fisheries, would enable source countries to control collection and trafficking of bio-genetic resources.[30] Individual transferable quotas are based on granting rights of harvest for a specific quantity and quality of the resource, but only for a specific duration. The advantage of this taxing method is that revenues can be generated for the conservation of the resources irrespective of their prospective use.

To sum up, it has been here argued that State sovereignty, although it is a property-allied concept, because of the exclusive right assigned to the sovereign to exercise control over a geographical region, is not identical to

[24] See German Bürgerliches Gezetzbuch (BGB 1900 and 1919) paras 1012–1017 and Italian Codice Civile (1942).

[25] Parisi (n 22) 21.

[26] UN Doc E/CN.4/Sub.2/2004/30, paras 43 and 44. [27] ibid.

[28] See *Awas Tingni v the Republic of Nicaragua: the Inter-American Court of Human Rights*, in interpreting the right to property as found in article 21 of the American Convention on Human Rights, made clear in its judgment that indigenous peoples' rights to their lands include rights to the resources there (para 153) and that these rights of ownership are held by the community in their collective capacity and according to their own customary law, values, customs and mores (paras 148, 151, 153). Though the Court did not use the term 'sovereignty', there is no question that the decision found that international law protects the governmental or collective right of the community to the land and resources.

[29] See *Alexkor Limited and the Government of South Africa v The Richtersveld Community*, Constitutional Court of South Africa—CCT 19/03 (14 October 2003), where the Court found that the indigenous Richtersveld Community not only had a right of communal ownership of the land under indigenous law, but also the right to use its water, its land for grazing and hunting and to exploit its natural resources, above and beneath the surface; the court concluded that ownership of the subsurface vested in the Community because ownership of subsurface resources was established by the indigenous law of the Community, and that these resources could not have belonged to anyone if they did not belong to the indigenous Community. Also *Delgamuukw v British Columbia of Canada*, 3 S.C.R. 1010 (1997), 153 D.L.R. (4th), 193 S.C.C. (Can. 1997); despite the limitations of the Canadian Supreme Court ruling to the aboriginal titles in Delgamuukw, the Court also made it clear that 'aboriginal title encompasses mineral rights' (para 122).

[30] P Gehl Sampath, *Regulating Bioprospecting: Institutions for Drug Research, Access and Benefit-Sharing* (UNU Press, Tokyo/New York, 2005).

state property.[31] States cannot really assign property rights over genetic resources; plant genetic resources exist beyond the border of one State. The role of the state in access and benefit-sharing contracts can thus not be *qua* owner of the genetic material; but *qua* supervisor of the bioprospecting undertakings within its territorial control, so that neither the resources are harmed nor the rights of its citizens impaired.

### *C. Intellectual Property Rights and Traditional Medicinal Knowledge*

A topical subject among academics and negotiators at international fora relates to intellectual property rights, traditional knowledge and bioprospecting. Two levels of the interface of intellectual property and benefit-sharing can be discerned: on the one hand the protection of the final product of commercial bioprospecting may qualify for intellectual property protection and there is a question on the distribution of the benefits derived from this protection; on the other hand there is the delicate discussion on the protection of traditional knowledge internationally through intellectual property instruments. The two levels are linked because international recognition of intellectual property over traditional knowledge will define the protection of the final product using this knowledge and hence the benefits' distribution. In this section an attempt is made to answer the following questions: is there a reason and benefit to design a new intellectual property instrument for the protection of traditional knowledge? If not, how can we protect this knowledge from abuse? Are there alternative ways to secure 'fair and equitable' benefit-sharing, once there is a commercial product that used biological resources and the associated traditional medicinal knowledge?

The question relating to the need for intellectual property protection of traditional knowledge is already in the centre of discussions at the WIPO Intergovernmental Committee and the TRIPS Council.[32] There is a need morally to recognize that traditional medicinal knowledge belongs to traditional cultural expressions, and, given that protection is far from adequate at the domestic and international levels, there is a further need to restrain others from commercializing it with no compensation at all to its holders. But does this moral obligation amount to the need for creation of a new intellectual property tool?

In Article 8j, the CBD specifically addresses the knowledge, innovations and practices of indigenous and local communities embodying traditional lifestyles relevant for the conservation and sustainable use of biological diversity. This provision is an acknowledgement of the contribution of traditional knowledge to the conservation and sustainable use of biodiver-

[31] See Chapter 3, Section 2.1a).
[32] See Chapter 4.

sity, but it offers no concrete protection. This topic has been covered extensively in scholar literature and the views vary considerably. The following paragraphs attempt to explore in a concise manner some traditional and non-traditional forms of domestic and international intellectual property rights, and whether and how their use could ease protection of traditional medicinal knowledge. Traditional forms of intellectual property rights are plant breeders' rights, trade secrets, patents, copyrights and geographical indications;[33] whereas in the non-traditional forms for intellectual property protection, there is a menu of other legal sources linked to human rights law, such as farmers' rights, folklore, cultural rights and intangible heritage protection, which hold some ancillary tenets for the protection to a variety of traditional cultural expressions.

### *1. Traditional forms of intellectual property protection and traditional knowledge*

Plant breeders' rights cannot accommodate traditional medicinal knowledge concerns, because of their exclusive relation to agriculture and horticulture; the applicability of UPOV 1961[34] for wild plants with possible medicinal uses is not a viable option, because it goes beyond the scope of the treaty.

Trade secrets provide for the confidential keeping of valuable information and the recovery of financial losses for improperly acquired information, but not when information was spread out because of careless or accidental loss of control or independent recovery. Since they actually limit the free information flow, trade secrets are not advantageous from a public perspective. For the protection of traditional knowledge, trade secrets could be used in the many cases where indigenous communities refuse the diffusion of medicinal knowledge for cultural, religious or other reasons. This conception, however, is problematic, because the rest of the world and people suffering from a specific ailment more in particular (indirect stakeholders as defined above), should be able to profit from this knowledge. Holders of such knowledge should thus be able to somehow add value, manufacture/elaborate/produce the drugs, or at least participate in the manufacture of the medicines and profit from their commercialization.

Copyrights and patents do not prove very functional for traditional knowledge protection either: both require the specific author of the work to be protected, whereas traditional knowledge is usually held by a community composed often by an undefined number of individuals; in

33 See also CJ Visser, 'Making Intellectual Property Laws work for Traditional Knowledge' in JM Finger and P Schuler (eds) *Poor People's Knowledge—Promoting Intellectual Property in Developing Countries* (OUP/The World Bank, Whasington DC, 2004) 241.

34 See also modifying acts of 1978 and 1991.

addition, both patents and copyright have a time limit, posing doubts about the time-limited protection of traditional knowledge. Although there are some exceptional cases of individual innovation, these tools fail to accommodate the dynamic process for the development of traditional knowledge. This is because copyrights, on the one hand, require 'fixation' of the work and do not recognize the frequently oral tradition through which traditional knowledge is recorded; and patents, on the other hand, are granted for a single act of invention, while traditional knowledge obviously fails to meet the patent requirements of inventiveness, novelty and industrial use.[35]

Finally, geographical indicators show the essential role of geographic and climatic factors, as much as of local peoples' knowledge, in the quality of certain products. The TRIPS-Agreement creates a two-level system: a basic protection in Article 22, according to which geographical indications have to be protected so as to prevent unfair competition practices and not to mislead the consumers; and the enhanced protection of its Article 23 for geographical indications for wines and spirits, in other words the *appellations d'origine*. Subject to a number of exceptions, the latter are to be protected even if misuse would not mislead the public. Member Countries enjoy a great degree of discretion in implementing the TRIPS provisions: they may chose among geographical indications, trademark law or business practices to protect the same subject matter, a situation that has created an amount of controversy within the TRIPS Council. In this study, it is argued that elaboration of the geographical indications' paradigm may serve as a good vehicle for the protection of traditional knowledge.[36]

### 2. *Non-traditional forms of intellectual property rights*

Among the non-traditional forms of intellectual property, farmers' rights, folklore, and cultural property treaties will be now examined.

'Farmers' rights' is a term developed by the FAO under the revised International Undertaking in an effort to respond to the complaints against Plant Breeders' Rights (PBRs). The problem with farmers' rights is that although they were conceived as counter-rights of PBRs, they have a completely different function from PBRs: while PBRs are intellectual property rights, farmers' rights operate more as a moral obligation and less as economic incentives, they are not connected with any specific action, and they constitute recognition of general conservation and equity objectives. Farmers' rights were supposed to be made operational through a compen-

[35] H Ullrich, 'Traditional Knowledge, Biodiversity, Benefit-Sharing and the Patent System: Romantics v Economics?' in F Francioni and T Scovazzi (eds) *Biotechnology and International Law* (Hart Publishing, Oxford, 2006) 202 f.

[36] See Section 2.3.c).

sation fund, but because of a lack of contributions, the fund has never been established. The concept of farmers' rights could be used for a similar protection of traditional knowledge holders within the CBD, however its failure to compensate the farmers anticipates that a similar institution under the CBD institutional framework would fail in the same way.

Also, the folklore protection presents many similarities to medicinal knowledge protection and with selected material—landraces protection: they are all the outcome of long-term community contribution and there is no compensation system or at least a sort of acknowledgement of these communities for their contributions. The problems one encounters in trying to protect medicinal products based on traditional knowledge through patents emerge also in the case of trying to apply copyrights and trademarks in folklore protection because folklore fails to meet the 'new and unique creations' requirement. For this reason, the WIPO specialized Intergovernmental Committee studies possible ways of protection for both folklore and traditional knowledge associated to genetic resources.

Actually, before forming the Intergovernmental Committee in 1999, WIPO tried to explore the contours of the traditional knowledge issue and the ramifications of intellectual property protection since the early 1980's and came up with the 'Model Provisions For National Laws On The Protection Of Expressions Of Folklore Against Illicit Exploitation And Other Prejudicial Actions'. These guidelines in the end remained unimplemented, because they had no political support, were vague, and failed to address the semantic and definitional difficulties of traditional cultural expressions, traditional knowledge and folklore.

Last but not least, cultural property treaties and tenets should be mentioned as non-traditional forms of intellectual property. In response to the plundering and destruction of many western artworks and monuments during the two World Wars, the 'Convention for the Protection of Cultural property in the Event of Armed Conflict' was adopted in The Hague in 1954,[37] with a view to providing for measures of protection to important works of the past for future generations. Later, in 1970, the UNESCO 'Convention on the Means of Prohibiting and Preventing the Illicit Import, Export and Transfer of Cultural Property'[38] established principles to regulate cultural property protection and stressed that every State should become aware of 'the moral obligations to respect its cultural heritage and that of other nations'.[39]

[37] Convention for the Protection of Cultural Property in the Event of Armed Conflict, The Hague 14 May 1954.

[38] Convention on the Means of Prohibiting and Preventing the Illicit Import, Export and Transfer of Ownership of Cultural Property, Paris 14 November 1970.

[39] ibid Preamble.

Furthermore, the UN Declaration on the Rights of Indigenous Peoples[40] calls for measures that protect and control their knowledge and practices.[41] But most importantly, UNESCO has gradually paved the way towards the understanding of cultural expressions and practices, their protection, and respect, with these efforts culminating with the adoption of the 'Convention for the Safeguarding of the Intangible Cultural Heritage',[43] on 17 October 2003, which came into force on 20 April 2006.[44] This Convention relies very much on a bottom-up approach for the safeguarding of cultural heritage, obliging its signatory states to take domestic measures to protect their own cultural heritage and then foster coherence and cooperation at the regional and international levels through the exchange of information, experiences and joint initiatives. Traditional medicinal knowledge could be included in the scope of the Convention's protection: very often such medicinal practices form part of the long-standing culture of indigenous peoples. The Convention prioritizes domestic to international action in this respect, something that should be considered by developing countries trying to introduce changes in the international patent system for the protection of such knowledge.

### 3.Protection of traditional knowledge from and for exploitation

The foregoing analysis shows that intellectual property forms as they stand today cannot accommodate the interests of traditional knowledge holders. Ten years ago, the system was heavily criticized for its inadequacy in the protection of such knowledge. Stemming from the environmental justice movement, which emphasized how environmental problems disproportionately harmed the poor, 'cultural environmentalism' expressed how intellectual property law harmed the poor and failed to recognize their cultural contributions, and suggested that preservation of the public domain as the solution.[44]

[40] See Economic and Social Council, Commission on Human Rights—Sub-Commission on prevention of Discrimination and protection of Minorities, *Discrimination Against Indigenous Peoples—Technical Review of the United Nations Draft Declaration on the Rights of indigenous Peoples, Addition: Draft Declaration as Agreed upon by the Members of the Working Group at its Eleventh Session*, Doc E/CN.4/Sub.2/1994/2/Add.1, 20 April 1994, Articles 24 and 29.

[41] See more analytically on the Declaration in Section 3.3.b).

[42] Convention for the Safeguarding of the Intangible Cultural Heritage, Paris, 17 October 2003.

[43] UN News Center, 'UN Treaty to Protect Oral Traditions enters into Force' (New York, 21 April 2006).

[44] J Boyle, *Shamans, Software, & Spleens: Law and the Construction of Information Society* (Harvard University Press, Cambridge, Mass, 1996); M Sunder, 'The Invention of Traditional Knowledge' (14 February 2006), UC Davis Legal Studies Research Paper No 75, published on the Internet at: http://ssrn.com/abstract=890657.

The problem is that in today's global society, the public domain offers inadequate protection of the traditional knowledge holders: if traditional knowledge remains in the commons, then it is commercialized by third parties, and the holders of traditional knowledge are the losers in such commercialization ventures. After all, holders of traditional knowledge have put great effort to familiarize themselves with Western intellectual property rights to learn, which they have started to use.[45] Indigenous groups, with the assistance of NGOs, file appeals against patents in defense of their traditional knowledge; whereas India, one of the greatest opponents of global intellectual property standards, has gradually abided by the TRIPS provisions.[46]

Ultimately, the goal is to protect traditional knowledge from misappropriation and at the same time to show to its holders how to benefit from it. Traditional forms of intellectual property rights could be either used in their present function to protect misappropriation, or accordingly be modified to accommodate traditional knowledge particularities.

The proposals to make i) the disclosure of source, origin or legal provenance, ii) the proof of a PIC certificate, and iii) the proof of fair and equitable benefit-sharing, compulsory requirements to grant a patent are more likely to fail, mainly because of the already mentioned debate at the international negotiations.[47] Such proposals would add unnecessary burden to legislature and administration, which would not be outweighed from the benefits. Also, past recommendations for the streamlining of PIC applications with a PIC internationally recognized certificate[48] have not been implemented.

Good use of the patent system as it stands today could provide for the defense of traditional medicinal knowledge. If the description of the source lines of the pharmaceutical invention is identical with traditional knowledge uses, then the patent should not be granted; or if such patent has already been granted, it should be invalidated. This practice is proven by patents recently revoked based on proof of 'prior art'.[49]

[45] See in Finger and Schuler (n 33) case studies 2–9.

[46] See also the Geographical Indication of Goods (Registration and Protection) Act 1999 (No 48 of 1999) (effective in 2003).

[47] See Chapter 4 and H Ullrich, 'Traditional Knowledge, Biodiversity, Benefit-Sharing and the Patent System: Romantics v Economics', in F Francioni and T Scorgzzi (eds) *Biotechnology and International Law* (Hart Publishing, Oxford, 2006).

[48] B Tobin, 'Certificates of Origin: A Role for IPR Regimes in Securing Prior Informed Consent' in J Mugabe et al (eds), *Access to Genetic Resources—Strategies for Sharing Benefits* (ACTS Press, Nairobi, 1997) 329.

[49] eg Reexamination Certificate, U.S. Patent No. 5,401,504 B1 (requested Oct. 28, 1996) (issued 21 April 1998) (canceling all claims in original patent on turmeric); Case No. T0146/01-3.3.2, Method for Controlling Fungi on Plants by the Aid of Hydrophobic Extracted Neem Oil (B.A. Eur. Patent Office Mar. 8, 2005) (upholding on final appeal the 2001 revocation of European Patent Office neem patent).

In the event that traditional knowledge contributes in the invention, which otherwise meets the criteria of novelty and industrial use, new ways for compensation of the holders of traditional knowledge should be devised, not related to the conditions to grant the patent. Indeed the treatment of traditional knowledge as know-how and the conclusion of a know-how license similar to software practice in the Peru ICBG was a very ingenious treatment of traditional knowledge.[50]

Many object that this is not a satisfactory solution: it entails high costs and standing issues before the user-countries' courts, and also because traditional knowledge holders have to be constantly alert, monitor, and bring law suits against unauthorized uses.[51] But, costs arise for any sort of other complain and litigation. In any event, it is easier to overcome procedural obstacles such as those associated to standing than to introduce substantive modifications in the patent law.

The non-traditional forms of intellectual property rights offer some interesting insights in promoting protection of traditional knowledge as 'cultural heritage'. There is a trend towards more holistic concepts to protect such knowledge.[52] A number of international treaties and non-traditional forms of intellectual property show that there is a need to recognize the cultural contribution of peoples;[53] the problem is that mere recognition does not satisfactorily cover the dynamism and inventiveness of the cultural contributions. The non-traditional forms of intellectual property do not challenge formal intellectual property instruments, but rather the utilitarian tenets for these instruments.[54]

[50] See Chapter 6 Section II.B.

[51] Ullrich in Francioni (n 47) 204 and CBD Doc UNEP/CBD/WG-ABS/3/5.

[52] See also C Antons, 'Traditional Knowledge and Intellectual Property Rights in Australia and Southeast Asia in C Heath and A Kamperman Sanders (eds) *New Frontiers of Intellectual Property Law: IP and Cultural Heritage, Geographical Indications, Enforcement, Overprotection* (Hart Publishing, Oxford, Portland, Oregon, 2005) 37.

[53] Also the CBD Article 8j and the Draft Access to Knowledge Treaty (also known as 'A2K Treaty') (9 May 2005). The Draft Treaty emerged from a move by Brazil and Argentina to develop the WIPO agenda and its goal involves facilitation of the transfer of knowledge to developing nations, and security of the viability of open innovation systems all over the world, see IP Justice, NGO submission to the WIPO Development Agenda, Publisher on the Internet at: http://www.ipjustice.org/WIPO/IIM3/NGO.Stmt.IIM3.Dev.Agenda.sp.pdf; 'El acceso al conocimiento y el intercambio de información son fundamentales para la educación e investigación y para fomentar la innovación y creatividad. Un tratado que establezca libertades para los usuarios contemplará 'la necesidad de mantener un equilibrio entre los derechos de los autores y los intereses del público en general, en particular en la educación, la investigación y el acceso a la información, como se refleja en el Convenio de Berna', tal como lo establece el Preámbulo del Tratado de la OMPI sobre Derecho de Autor. Un tratado sobre acceso al conocimiento y la tecnología sería un componente clave de las políticas para aliviar la situación de países en desventaja y beneficioso para el desarrollo integral socioeconómico y político de un país.'

[54] Sunder (n 44) 23.

The recognition of the cultural heritage and moral rights of indigenous societies and their members pave the way for the design of mechanisms aiming at compensating indigenous peoples for their creative power and inventive minds.[55] Indeed, intellectual property can be well founded in terms of dignity and human rights: the rationale of the Berne Convention[56] is based on the civil law countries personality-derived *droit d'auteur*, and not to the strictly economic criteria for copyright according to the common law tradition.[57] Clearly, the inalienable, universal, and non-seizable moral rights of the authors triggered the formation of an intellectual property tool that yields economic benefits.

There is a certain risk of commoditization of culture and knowledge that already belongs to the commons; it is very difficult to draw the dividing line between cultural elements that generally belong to the commons, on the one hand, and those that have the potential to be commoditized. Even if the commercialization of cultural elements may raise criticisms,[58] the fact is that cultural goods have always been marketed through human history, and they will continue being marketed. In the end, trade and cultural stakes are not necessarily at odds. The problem is that cultural heritage protection alone cannot easily meet the market demands. To date, the fact that medicines derive from contributions of traditional knowledge and are commercialized, demonstrates that traditional knowledge is a cultural element being de facto expressed in commercial terms; and it is not adequate to recognize that such knowledge belongs to its holders; it is also to compensate them in commercial terms.

The history of intellectual property law shows that the list of intellectual property tools is not exclusive; new tools have been devised to accommodate new stakes, when these did not fit in the existing tools. Especially the 20th century proved prolific in intellectual property, as numerous new rights

[55] Human rights discourse considerably challenges the thinking of intellectual property; see LH Helfer, 'Toward a Human Rights Framework for Intellectual Property' 40 (2006–2007) UC Davis Law Review Symposium Issue (2006–2007).

[56] See Berne Convention for the Protection of Literary and Artistic Works, Berne, 24 July, 1971, as amended on September 28, 1979, especially Article 6bis: (1) independently of the author's economic rights, and even after the transfer of the said rights, the author shall have *the right to claim authorship of the work and to object to any distortion, mutilation or other modification of, or other derogatory action in relation to, the said work, which would be prejudicial to his honor or reputation* (*emphasis* added). This could serve as a basis for the protection of folklore and other traditional artistic expressions of a group or community.

[57] Because of this conceptual difference and the denial of moral rights in the Anglo-Saxon world, the US resisted for more than hundred years to join the Berne Convention; however the Visual Artists Rights Act of 1990 (VARA) is the US adoption of the moral rights concept into codified law.

[58] MF Brown, *Who Owns Native Culture* (Harvard University Press, Cambridge, Massachusetts, 2003) 94 f and 173 f: 'Indigenous representatives all over the world speak as *owners* of their culture, in order to watch and assert control over the cultural property use from non-indigenous peoples'.

appeared next to the traditional patents, copyrights and trademarks, such as plant breeders' rights or integrated circuits and database protection (especially in the EC).[59]

In fact, international intellectual property standards were established in an environment of global power relationships, where concerns and needs of developing countries were not really reflected. Until very recently, peoples' different cultures were not considered as being of the same value, and were even ranked progressively from 'savagery' to 'barbarism' to 'civilization'.[60] As a result, 'hierarchies of culture' defined how decisions were taken in most areas of State cooperation, excluding 'less civilized' developing and tribal societies from decision-making. Intellectual property law was also designed without due consideration to the needs and stakes of indigenous cultures, because such cultures were seen as less developed.

Because of these sad origins of international intellectual property standards, the system of values sustaining current intellectual property law is completely different from the system of values of indigenous peoples. This 'Western' sense of intellectual property does not reflect the incentives for creation and the appreciation of creation, as conceived by indigenous peoples. Indigenous peoples do not see cultural property as 'property', but as an essential expression of their culture.[61]

Presently, the increasing protection of human rights promotes equality of different cultures and civilizations of mankind and the need to accommodate interests for the protection of traditional knowledge. It is not the protection of this knowledge that is argued here. Quite the contrary: the need for protection of traditional knowledge and the compensation of its holders is advocated. What is challenged is the creation of an additional intellectual property tool, when the system allows for improvements that protect such knowledge from abuse. Can a new international norm be flexible enough to encompass the customary legal diversity of traditional societies? Can such a new norm be very different from the existing system? And do the benefits of additional criteria in patent law, for instance, outweigh the costs associated?

The creation of a *sui generis* protection for traditional knowledge has

[59] See Directive 96/9/EC of the European Parliament and of the Council of 11 March 1996 on the legal protection of databases.

[60] OB Arewa, 'TRIPS and Traditional Knowledge: Local Communities, Local Knowledge, and Global Intellectual Property Frameworks' (2006) 10 Marquette Intellectual Property Law Review 159.

[61] See also the Australian cases *Yumbulul v Reserve Bank of Australia* (1991) and *Milpurrurru v Indofurn Pty Ltd* (1995) and *Bulun Bulun v R & T Textles Pty Ltd* (1998), cited by Torsen (2005). M Torsen, 'Anonymous, Untitled, Mixed Media: Mixing Intellectual Property Law with Other legal Philosophies to Protect Traditional Cultural Expressions' (2005) 12 *CASRIP Newslettter-Spring/Summer 2005*, published on the Internet at: https://www.law.washington.edu/Casrip/Newsletter/Vol12/newsv12i1Torsen.html (23/12/08).

been extensively discussed by academics[62] and within WIPO.[63] Shaping a new intellectual property tool for the exclusive protection of traditional knowledge does not appear a very good idea, not only because this option is not pragmatic at the moment, given the controversies surrounding the *sui generis* protection for traditional knowledge at the negotiating tables, but also for reasons relating to the substance of such a right and its capacity to actually benefit the traditional knowledge holders and the international community more in general:[64] there is a vast array of traditional knowledge categories[65] and there is a significantly difficult task to define the individuals constituting the community/beneficiary,[66] implying additional costs and expenses. Even the WIPO Intergovernmental Committee, which is currently studying the creation of such a right, observed that

> no single solution is likely to meet all the needs of indigenous and traditional communities and to protect traditional cultural expressions in a comprehensive manner. Instead, effective and comprehensive protection may be found in a 'menu' of options, comprising differentiated and multiple levels and forms of protection.[67]

In an effort to reconcile public and private interests in intellectual property policy, the application of the open source or free software movement[68]

[62] Such as Traditional Knowledge Property Rights (TIP-Rights) or Traditional Resource Rights (TRRs), see A von Hahn, *Traditionelles Wissen Indigener und Iokaler Gemeinschaften zwischen Geistigen Eigentumsrechten und Public Domain* (Springer Verlag, Berlin, Heidelberg, New York, 2004; T Cottier and M Panizzon, 'Legal Perspectives on Traditional Knowledge: The Case for Intellectual Property Protection' (2004) 7 Journal of International Economic Law 371; Biber-Klemm in T Cottier and PC Mavroidis (eds), *Intellectual Property—Trade, Competition and Sustainable Development* (University of Michigan Press, Ann Arbor, 2003) 481; T Cottier, 'The Protection of Genetic Resources and Traditional Knowledge: Towards More Specific Rights and Obligations in World Trade Law (1998) 1 Journal of International Economic Law 555; D Posey and G Dutfield, *Beyond Intellectual Property: Towards Traditional Resource Rights for Indigenous Peoples and Local Communities* (International Development Research Centre, Ottawa, 1996); A Subramanian, 'Genetic Resources, Biodiversity and Environmental Protection' (1992) 26 Journal of World Trade 105.

[63] See Chapter 4 Section VI.

[64] See critique by Ullrich in Francioni (n 17); also D Posey and G Dutfield, *Beyond Intellectual Property: Towards Traditional Resource Rights for Indigenous Peoples and Local Communities* (International Development Research Centre, Ottawa, 1996).

[65] See also A Choudry, 'Corporate Conquest, Global Geopolitics: Intellectual Property Rights and Bilateral Investment Agreements' Seedling, January 2005, 7; P Kuruk, 'Bridging the Gap between Traditional Knowledge and Intellectual Property Rights: Is Reciprocity an Answer? (2004) 7 The Journal of World Intellectual Property 429.

[66] In the ICBG Peru, for instance (see Chapter 6 Section II.B.), there was a serious problem in the representation of the indigenous people in the first and second set of negotiations and still its legitimization remains controversial

[67] WIPO—IGC, *Traditional Cultural Expressions/Expressions of Folklore, Legal and Policy Options*, Sixth Session, Doc. WIPO/GRTKF/IC/6/3, 1 December 2003, para 2.

[68] Like Linux, Apache or Mozilla or the online encyclopedia Wikipedia. See also Institute for

has been suggested in areas outside the digital world, such as biology and pharmaceuticals.[69] Open source describes practices in production and development that promote access to the end product's sources in a way that the users own the patents to their creations, but cannot hinder others from using the original shared information to develop similar products. As an alternative production method, it involves a great amount of both information- and decision-making sharing and allows democratic participatory mechanisms in intellectual property policy-making.[70]

The development of medicines from traditional knowledge could benefit from the open source production paradigm: it would involve the design of websites, programs and databases, where drug candidates are available to any individual or company that is interested in developing them, and contracts are then signed with the company offering the best terms. The problem is that this model would require immense capacity on computer software and technology use on the part of traditional knowledge holders. Also, there would be a specific supervision problem, as once on the website, traditional knowledge can be easily accessed and illegally used by anyone and any company.

Another option promotes the potential benefits of compensatory liability rules over property rules[71] putting emphasis on the public domain in the technological universe.[72] Liability rules entitle investors to receive money to

OneWorld Health at http://www.oneworldhealth.org/, Drugs for Neglected Diseases Initiatives at http://www.dndi.org and Tropical Disease Initiative at http://www.tropicaldisease.org.

[69] MEI Kipp, 'Software and Seeds: Open Source Methods' (2005) 10 First Monday, published on the Internet at: http://www.firstmonday.org; SM Mauer, A Rai and A Sali, 'Finding Cures for Tropical Diseases: Is Open Source an Answer?' (2004) 1:e56 PLoS Med, 183.

[70] Even if conceptualized as an independent and opposing system to intellectual property rights, open source is consistent with property theory and builds upon an environment of commercial importance, like software or (potentially) biotechnology, and in the end it is just another type of property in R&D. In today's economic system large firms in software and biotechnology can exploit commercial interactions with open source in a much better way that small firms, leading to concentration; so whatever the property mechanism selected, its structure and content will inevitably portray the structure of the economic system. See relevantly the critique by R Mann, 'The Commercialization of Open Source Software: Do Property Rights Still Matter?'(2005) U of Texas Law, Law and Econ. Research Paper No 58; Parisi (n 22); DJ Faye, 'Bioprospecting, Genetic Patenting and Indigenous Populations—Challenges under a Restructured Information Commons' (2004) 7 Journal of World Intellectual Property 419.

[71] There are two broad theories of property, 'intertranslatable and compatible': the bundle-of-rights analysis viewing property as a package of rights among persons with respect to things, developed by Hohfeld (1919) and then by Honoré (1961); and the rule-governed entitlements analysis proposed by Calabresi and Melamed (1972), according to which 'an entitlement is, roughly, an interest that the law does or should protect' ... 'by using one or more different sorts of rules: property rules, liability rules and rules of inalienability'; see Muzner in MP Golding and WA Edmundson (eds) *The Blackwell Guide to the Philosophy of Law and Legal Theory* (Blackwell Publishing, Oxford, 2005).

[72] Visser in Finger and Schuler (n 33) 231; RN Nwabueze, 'Symposium: Traditional Knowledge, Intellectual Property, and Indigenous Culture Articles, Ethnopharmacology, Patents and the Politics of Plant Genetic Resources' (2003) 11 Cardozo Journal of International and Comparative Law 585; MF Brown, *Who Owns Native Culture?* (Harvard

pay their R&D costs from second comers who freely borrow the resulting sub-patentable know-how for their industrial applications for a specified period of time. In access and benefit-sharing, liability rules would allow the involuntary transfer of an entitlement, on the condition that an objectively determined value is paid by the receiver and a State agency performs the whole function, as a version to implement a compulsory licensing scheme.[73] The problem of this proposal for a globalized compensation system is that the diversified expectations of the stakeholders, circumstances, and needs and values of indigenous peoples seriously challenge its implementation.[74]

Otherwise, the formation of a Global Biocollection Society possibly under the auspices of the World Bank has been suggested,[75] for the protection of traditional ecological knowledge and genetic resources. Companies and groups with claims to traditional economic knowledge would become members, and the Society could assist in negotiations between stakeholders, set contracts standards, and provide a dispute settlement system. The advantages of such a repository system are that genetic resources and associated knowledge can become profitable independently of the state of play of international negotiations for a universal access and benefit-sharing treaty or the formation of a *sui generis* intellectual property tool; in addition, dealing with a single mediating mechanism instead of with numerous national bureaucracies, diminishes transaction costs.

Such a repository mechanism could actually be created under the auspices of the CBD Secretariat instead of the World Bank and would have the ability to warrant fair and equitable terms in access and benefit-sharing undertakings. In addition, it is here suggested that for the protection of traditional knowledge, the model provided by geographical indications and especially by the French concept of *appellation d'origine* offer a useful analogy for traditional knowledge and traditional expressions: this model provides protection for *some* types of local knowledge. The extension of the scope of Article 23 is rightly sought by countries such as India, which have many world-renowned geographical indications, such as Darjeeling tea and Basmati rice, but lack international higher protection.[76]

Indeed, the land features in conjunction with traditional contributions by humans in the final product could fit in a template for respect, protection and

University Press, Cambridge, Mass, 2003); JH Reichman, 'Of Green Tulips and Legal Kudzu: Repackaging Rights in Subpatentable Innovation' (2000) 53 Vanderbilt Law Review 1743.

[73] Brown (n 72) 239; if radio stations were required to obtain prior permission before broadcasting, transaction and broadcasting costs would be very high.

[74] Nwabueze (72).

[75] P Drahos and J Braithwaite, *Information Feudalism: Who Owns the Knowledge Economy?* (New Press, New York, 2003); Visser in Finger and Schuler (n 33) 232.

[76] See WTO site on Geographical Indications at: http://www.wto.org/english/tratop_e/trips_e/gi_e.htm; Arewa (n 60) 161; Sunder (n 44) 18.

recognition of indigenous communities' medicinal knowledge and practices similar to the respect for the contribution of winegrowers and breeders (plant breeders' rights). Such protection could be awarded though only in cases where the final product is identical to its local use and its value has not been altered through significant technological and scientific additions; if not, the new medicinal product may qualify for patent protection and other means of compensation should be sought.

Hence, domestic geographical indications in combination with the revocable character of patents, in the case of misappropriation of traditional knowledge, can protect traditional knowledge in a two-fold manner: defending and promoting use and development. On the one hand, when traditional knowledge over the medicinal qualities of plants formed the basis for medicines, a patent cannot be granted or if granted, it can be cancelled, because of 'prior art' proof; today, the actual problem stems more from the narrow 'prior art' definition by domestic laws[77] or from a lack of adequate information at the patent offices. On the other hand, expansion of the scope of the geographical indications and the creation of registries, not only serves 'prior art' research, but if the holders learn how to use these tools, they can profit from their knowledge and its applications.

National efforts to register and document traditional knowledge have recently taken place, especially India,[78] and followed by Venezuela (Biozulua).[79] Such registration and inventory systems have been criticized for possible adverse effects, such as the need to define access to information, the required control, and registering knowledge in the name of third parties.[80] However, the case of India, which to date presents the most elaborate system, appears to be rather successful and promising, as more and more people register their knowledge. In the end, organization and publication of traditional knowledge data for public use fosters research, whereas prior publication renders the efforts of those who would later seek to patent this data unsuccessful, as they cannot meet the novelty require-

[77] US Act, Title 35, Patents, Part II, Chapter 10, §102.

[78] The Geographical Indication of Goods (Registration and Protection) Act, 1999 No. 48 of 1999 (effective in 2003); J Lancaster, India folk remedies protected from piracy by drug companies, Washington Post, Sunday, 15 January 2006; NS Gopalakrishnan, 'Protection of Traditional Knowledge—The Need for Sui Generis Law in India' (2002) 5 The Journal of World Intellectual Property 790.

[79] V Eugui, 'El régimen legal y la experiencia venezolana en materia de acceso a los recursos geneticos, los conocimjentos traditionales y la propiedad intelectual' DPI & Mercado, Materia de la carrera de Abogacía en la Universidad de Buenos Aires, Profesor a cargo dra T Zamudio, published on the internet at: http://www.dei.bioetica.org/dpcdi.vivas.htm (16/6/09); this system was also suggested in the Peruvian Regime Proposal for the protection of the Collective Knowledge of the Indigenous Peoples and Access to Genetic Resources.

[80] M Alexander et al, *The Role of Registers and Databases in the Protection of Traditional Knowledge: A Comparative Analysis*, United Nations University/Institute of Advanced Studies Report, (UNU/IAS, Tokyo, 2004); J Cabrera-Medaglia, Costa Rica: Legal Framework and Public Policy in Carriroza et al (eds) (2004) 117; Gopalakrishnan (n 78).

ment. The recent UN Convention on intangible heritage also calls for national inventories of cultural elements and traditional medicinal knowledge, which are considered part of a culture and their registration would help prior art research. And WIPO has set up the WIPO Portal of Traditional Knowledge to help users find and use such knowledge.[81]

There is clearly a need to protect the cultural heritage of indigenous peoples, and respect the moral rights and dignity of the holders of traditional knowledge, from abuses relating to the functioning of the global market and the commercialization of cultural goods. This study argues that protection should primarily start at the domestic level, with laws that guarantee indigenous peoples' rights on intangible resources, as the 'Convention for the Safeguarding of the Intangible Cultural Heritage' mandates; then the discussion can move more dynamically to the international level.

In support of this argument, a WIPO study in 2002 on 'National Experiences with the Protection of Expressions of Folklore/Traditional Cultural Expressions: India, Indonesia and the Philippines' highlighted the need for national recognition at the domestic level that communities and tribes are entities within the territory and that domestic legislation should allow for 'gap fillers' for tribal law. The report indicated that the Philippine's IPRA is rather successful in this regards because it provides a lot of room for the implementation of the community customary laws.[82]

## III. INDIGENOUS PEOPLES' RIGHTS

Indigenous peoples' rights are in the center of current international discussions; issues such as discrimination of indigenous peoples within states, their representation, and the claims over their ancestral lands and their cultural heritage have received particular attention. At the United Nations Conference on Environment and Development in 1992 the need that indigenous peoples rights be included in the environmental agenda was highlighted,[83] whereas on 10 December 1994, the United Nations General Assembly officially launched the International Decade of the World's Indigenous People.[84]

[81] Portal of Online Databases and Registries of Traditional Knowledge and Genetic Resources at: http://www.wipo.int/tk/en/databases/tkportal/.

[82] PV Kutty and G Valsala, National Experiences with the Protection of Expressions of Folklore/Traditional Cultural Expressions: India, Indonesia and the Philippines, A Study on the Protection of Expressions of Folklore for the World Intellectual Property Organization (WIPO), November 25, 2002, published on the Internet at: http://www.wipo.int/tk/en/studies/cultural/expressions/study/kutty.pdf.

[83] Agenda 21 Chapter 26.

[84] International Decade of the World's Indigenous People, GA Res A/RES/49/214 of 23 December 1994.

Although human rights instruments have been extensively used in support of indigenous claims,[85] there is no agreement on the conceptual categorization of indigenous peoples' rights. Should indigenous peoples be treated as minorities? Should indigenous peoples' issues be addressed within the existing framework of international human rights law? Or by contrast, there is a need for recognition of a new legal category specifically on indigenous peoples' rights?[86] Although universal human rights and the existing minority rights can operate in the protection of indigenous peoples' rights, one of the objectives of indigenous peoples' movement is the construction and affirmation of a distinct program.[87]

It is not the objective of the current thesis to examine this unresolved problem in international law on the treatment of indigenous peoples' rights. What follows is a concise demonstration and analysis of the current human rights instruments and other international legal norms that speak for the indigenous peoples' rights in the access and benefit-sharing process foreseen in the CBD. Although, at first glance, there is no direct linkage between human rights and access and benefit-sharing, it is argued that there is a reciprocal relationship between the two areas: on the one hand the development and implementation of an access and benefit-sharing regime under the CBD may have a positive or negative impact on the respect for and exercise of the right embodied in human rights instruments, particularly with regard to the protection of the indigenous/local communities' rights; on the other hand, the strengthening of indigenous peoples' rights at the international level will impact considerably on the progression for international and domestic access and benefit-sharing legislation.

A rights-based approach in the access and benefit-sharing regulation aligns it to the CBD objectives for fair and equitable benefit-sharing and ultimately for sustainable use and conservation. More specifically, in the medicinal biodiscovery discussion, indigenous peoples' rights come into play with regards to two property-related fronts, tangible and intangible rights.

The first concerns indigenous peoples' rights over the lands and resources they manage; when these rights are respected at the domestic

[85] J Castellino and N Walsh (eds) *International Law and Indigenous Peoples* Raoul Wallenberg Institute Human Rights Library Series (M Nijhoff, Leiden, Boston, 2005; SJ Anaya *Indigenous Peoples in International Law* (Oxford University Press, Oxford, 2004); S von Lewinski (ed) *Indigenous Heritage and Intellectual Property—Genetic Resources, Traditional and Folklore* (Kluwer Law International, The Hague, London, New York, 2004); P Thornberry, *Indigenous Peoples and Human Rights* (Manchester University Press, Manchester, 2002) 116–241; P Aikio and M Scheinin (eds) *Operationalizing the Right of Indigenous Peoples to Self-Determination* (Institute for Human Rights, Abo Akademi University, Turku, Abo, 2000); S Pritchard (ed) *Indigenous Peoples, the United Nations and Human Rights* (Zed Books Ltd, London, 1998).

[86] B Kingsbury, 'Reconciling Five Competing Conceptual Structures of Indigenous Peoples' Claims in International and Comparative Law' (2002) 34 International Law and Politics 101.

[87] ibid.

level, indigenous peoples are recognized as direct stakeholders in the bioprospecting undertakings, they are negotiating actors and grant the PIC for the resources to be studied and commercialized. The recognition of indigenous peoples' rights over the lands and resources they manage is directly linked to current discussions on the permanent sovereignty of indigenous peoples' over their lands and recognition of their customary proprietary systems.

The second property-related front concerns medicinal traditional knowledge, the protection of the knowledge of indigenous communities associated to the resources, which can contribute to modern drug development.[88] The debate for the protection of traditional knowledge in the age of global knowledge has already been discussed.[89] This section intends to underscore that the protection of indigenous peoples' rights and values as enunciated at the international level can be primarily materialized by legal and policy action at the domestic level; such domestic action will promote indigenous peoples' tangible and intangible rights and will then give rise to a more comprehensive international protection.

This separation between tangible and intangible rights of the indigenous peoples is only symbolic, because in effect it is difficult to detach the concept of indigenous peoples' relationship with their lands, territories and resources, from the concept of their values system, culture, and developed knowledge and practices.[90] In the terms of the Mataatua Declaration, the outcome of the 'First International Conference on the Cultural & Intellectual Property Rights of Indigenous Peoples', the world's indigenous peoples were declared to 'have the right to self-determination and in exercising that right must be recognized as the exclusive owners of their cultural and intellectual property'.[91]

### A. Indigenous Peoples' Property Rights over the Resources

The task assigned to the States to legislate on access and benefit-sharing legislation under terms that meet CBD objectives becomes a complex issue

[88] See also J Gibson, 'Intellectual Property Systems, Traditional Knowledge and the Legal Authority of Community' (2004) European Intellectual Property Review 280.

[89] See Section II.C of this Chapter.

[90] The HRC interpreted the notion of *culture* as affording protection to the nature-based way of life and economy of indigenous peoples, see M Scheinin, 'Indigenous Peoples' Rights under the International Covenant on Civil and Political Rights' in Castellino and Walsh (eds) (n 85) 5; E-I A Daes, 'The Spirit and Letter of the Right to Self-Determination of Indigenous Peoples: Reflections on the Making of the United Nations Draft Declaration' in Aikio and Scheinin (n 85) 81.

[91] Commission on Human Rights, Sub-Commission of Prevention of Discrimination and Protection of Minorities, Working Group on Indigenous Populations (19–30 July 1993), First International Conference on the Cultural & Intellectual Property Rights of Indigenous Peoples Whakatana (12–18 June 1993) Aotearoa, New Zealand.

when it comes to ownership, land tenure and use of resources and access to them in local/indigenous lands. States that fail to recognize indigenous peoples' rights fall short of the CBD requirements in their access and benefit-sharing regulation.

Respect for indigenous peoples' property rights over their lands and resources derives from the recognition of their permanent sovereignty of indigenous peoples over these lands and resources. By accepting indigenous peoples' permanent sovereignty over natural resources, States engage to recognize ownership, management or control systems that indigenous communities developed, which differ from the property rights institutions as developed by Western societies.

In turn, the recognition of the permanent sovereignty over their lands and resources stems from recognition of the right to self-determination. Self-determination was primarily conceptualized as a political aspiration against colonization and gradually crystallized into a principle and a right. The right to self-determination would become defunct, lacking ownership, management, and control of the natural resources in the territory.

A number of States with indigenous populations, like the United States and Canada, introduced a system of trusteeship for the management by the State of indigenous resources, which led to the formulation of the trusteeship doctrine in international law and its 'civilizing mission'.[92] These systems have been criticized as reminiscent of the abuses of the past century's colonization system and charges of mismanagement have been recorded.[93]

Presently, there is a need for trusteeship to revive and for self-determination to move away from any fear relating to ceding sovereignty; there is a need that these two concepts move towards broader notions of human dignity and solidarity so as to allow indigenous stewardship over resources to develop.[94] Given that because of the ideology of the past century attempts to impose national policies on Indian reservations even with the objective of 'conservation' failed, it should eventually be acknowledged that allowing for more expansive exercise of indigenous property rights is more likely to promote sustainability.[95]

Therefore, States that generally provide for the protection of the indigenous and local populations in their territories should ensure in their access and benefit-sharing legislation that the customary land tenure systems of those responsible for the protection and sustainable use of the resources must be respected.

[92] Anaya (n 85) 31 f.

[93] Daes (n 90) UN Doc E/CN.4/Sub.2/2004/30, para 37.

[94] I Sutton and RL Clow (eds) *Trusteeship in Change: Toward Tribal Autonomy in Resource Management* (University Press of Colorado, Boulder, 2001).

[95] ibid.

### B. Self-Determination and Permanent Sovereignty over Natural Resources

The traditional principle of permanent sovereignty is a central principle of international law and of the system of the nations. It dates back to the 16th century that sovereign countries are 'territorial polities' that are *legally equal, differing in capabilities ... but with the same standing in international society, which means that the norm of non-intervention is central—no sovereign has the right to intervene in the affairs of another*.[96] This implies that independent states have an inalienable (*permanent*) right of sovereignty over their territories and the natural resources existing within them.[97]

In 1962, the General Assembly declared the right of *peoples* and nations (emphasis added) to permanent sovereignty over their natural wealth and resources, and affirmed that this right must be exercised in the interest of national development and of the well-being of the *people and the State* concerned (emphasis added).[98] By doing so, the General Assembly actually accelerated the procedure towards 'crystallization' of the concept of permanent sovereignty of indigenous peoples over natural resources as a legal principle in 1966 with the UN International Covenants on Human Rights, as will be shown later on in this chapter.

In recent years, self-determination is indeed one of the most contentious ideas in the international community. The classical concept of sovereignty in international law was not initially conceived to encompass the sovereign rights of indigenous peoples within sovereign states. Today, self-determination is the summa or synthesis of individual human rights and vice-versa: respect for human rights and self-determination supplement and strengthen each other.[99]

The strong indigenous peoples' and human rights movement pressed forward not only the recognition of the existence of more than one 'peoples' within the territory of the country, but also the notion of self-determination as a community human right,[100] enjoyed by States and 'peoples' within the State. The right of self-determination is still rather weak, but it is the

[96] G Dutfield, *Intellectual Property, Biogenetic Resources and Traditional Knowledge* (Earthscan, London, 2004) 5; C Brown, *Sovereignty, Rights and Justice* (Blackwell, Malden, MA, 2002) 35.

[97] See also K Hossain, 'Introduction' in K Hossain and S Roy Chowdhury (eds) *Permanent Sovereignty over Natural Resources In International Law: Principle and Practice* (Frances Pinter, London, 1984) ix.

[98] Permanent sovereignty over natural resources, GA Resolution 1803 (XVII), 1194th plenary meeting, 14 December 1962.

[99] A Cassese, *Self-Determination of Peoples—A Legal Reappraisal* (Cambridge University Press, Cambridge, 1995) 337.

[100] J Castellino, 'The "Right" to Land, International Law & Indigenous Peoples' in Castellino and Walsh (eds) (n 85) 107 f; R McCorquodale (ed) *Self-Determination in International Law*, (Ashgate, Dartmouth, Aldershot, Burlington USA, Singapore, Sydney, 2000) xi f; A Cassese, *International Law* (Oxford University Press, Oxford, 2005).

primary means through which indigenous rights to territory can be expressed. Hence, we can speak of major advances in international thought and action on indigenous issues and rights, as colonized peoples' human rights struggles to achieve political and economic self-determination gave rise to the principle of permanent sovereignty over natural resources in modern international law.

### *C. Indigenous Peoples in International Treaty Law*

Until very recently, indigenous peoples were seen as an obstacle to development. It was only in the early 1970s that indigenous groups directly concerned made the situation change. Today, despite the doubts on the definition of 'indigenous',[101] indigenous peoples constitute a unique issue in modern international law; they represent one of the most active lobbying interests in international organizations and they have managed to drive the adoption of several conventions and internal policies at international development agencies. Nevertheless, the idea that indigenous peoples restrain development persists in some parts of the world.[102]

#### *1. The International Labour Organization*

The International Labour Organization (ILO) was the first international agency working directly on the situation of indigenous peoples.[103] The interest of the organization on indigenous peoples was the natural outgrowth of its work on labor and social policy.

As early as 1921, the ILO began work on the situation of 'native workers' in the European colonies that led to the Forced Labour Convention in 1930 (ILO Convention No 29),[104] a Convention coping with the situation of dependent peoples who face pressure and even assimilation from external cultures. Later, the ILO had the leading role at the UN 'Andean Indian Programme', an integrated program for regional development of several countries and their indigenous peoples. In the framework of this program, the ILO examined the situation of indigenous peoples worldwide and in 1953 it published a detailed study cataloguing the living and working conditions of indigenous peoples around the world.[105]

[101] Thornberry (n 85) 12 f.

[102] L Swepston, 'Indigenous Peoples in International Law and Organizations' in Castellino and Walsh (n 85) 53.

[103] For background information on ILO work with indigenous and tribal peoples see ILO, Indigenous and Tribal Peoples, Background on ILO work with indigenous and tribal peoples, at http://www.ilo.org/public/english/indigenous/background/; Thornberry (n 85) 320 f.

[104] International Labour Organization No. 29 (C29), Forced Labour Convention, 1930, adopted on 28 June 1930.

[105] Swepston in Castellino and Walsh (n 85) 54.

The ILO Convention No 107 (ILO C107) on Indigenous and Tribal Populations in 1958[106] and later the ILO Convention 169 (ILO C169) concerning Indigenous and Tribal Peoples in Independent Countries in 1989 are the only comprehensive international binding instruments of the rights of indigenous peoples and of the States' obligations towards them. These two ILO Conventions cover a wide range of subjects and are not merely 'labour' conventions, like for instance other ILO Conventions which could be used for the protection of indigenous peoples in their capacity as workers.[107]

ILO C107, the first international convention on the subject of indigenous peoples on behalf of the entire UN system, covers basic policy and administration, protects customary law, and provides for special measures in matters concerning labor, social security, health, vocational training and general education for the achievement of equal treatment. Articles 11 and 12 of ILO 107[108] are critical for land rights protection: they recognize respectively the right of collective or individual ownership of the populations over the lands they occupy[109] and assert the obligation of the states to obtain the free prior consent of the indigenous populations before their removal from the habitual territories, which may occur only in a legal manner and in exceptional cases related to national security, national economic development and health of the populations.[110]

The flaw of ILO 107 was that it reflected the general belief at the time that indigenous societies were 'backward' and temporary and that for their survival they had to be brought into the national mainstream through integration and assimilation. For this patronizing and integrationist approach, ILO 107 received a lot of criticism.[111] Nevertheless, ILO 107 had an immense positive influence at both national and international levels: it drew international attention upon the claims for the survival and well-being of indigenous peoples and, as a result, policy actions were initiated. Without ILO 107, ILO 169 might have never been adopted. ILO 107 continues to be in force in countries, like India, which have not signed its successor, ILO 169.[112]

[106] International Labour Organisation Convention No 107 (C107) on Indigenous and Tribal Populations, adopted on 26 June 1957.

[107] See ILO No 29 and 1957 ILO Convention No. 105 on Abolition of Forced Labour; 1958 ILO Convention No 111 against Discrimination in Employment and Occupation and 1973 and 1999 ILO Conventions No 138 and No 182 respectively against child labour.

[108] Under PART II Land.

[109] ILO 107 Article 11.

[110] ibid Article 12.

[111] Swepston in Castellino and Walsh (n 85) 55–56; International Labour Organisation, ILO Convention on Indigenous and Tribal Peoples, 1989 (No.169): A Manual, Project to promote ILO policy on indigenous and tribal peoples (International Labour Office, Geneva, 2003) 4.

[112] Swepston in Castellino and Walsh (n 85) 56.

As a consequence, ILO 169, which entered into force on September 5, 1991, came to address the shortcomings of ILO 107. It revises ILO 107, and marks a change in the ILO's approach to indigenous and tribal peoples; protection is still the principal objective, but it is based on respect for indigenous and tribal peoples' cultures, distinct ways of life, and their traditions and customs. Indigenous and tribal peoples have the right to continue to exist with their own identities and the right to determine their own way and pace of development,[113] while they should 'participate in the formulation, implementation and evaluation of plans and programmes for national and regional development which may affect them directly.'[114] In addition, their customary laws are to be taken into account in the national laws application.[115]

The special cultural and spiritual relationship of indigenous peoples to their lands and territories is emphasized together with the collective character of this relationship.[116] The Convention clearly states that indigenous and tribal peoples have ownership and possession rights to the land they traditionally occupy;[117] hence, governments must take steps as necessary to identify such lands and assign such rights.[118]

It further specifies that the rights of the peoples to the natural resources pertaining to their lands must be safeguarded, including the right to participate in the use, management and conservation of these resources.[119] Finally, the States' rights to retain ownership of mineral or subsurface resources within indigenous lands is defended under the condition that it is exercised after consultation with indigenous peoples and that these peoples participate in the benefits and receive compensation for any damages.[120]

The rights of indigenous peoples to participate in the negotiating process, grant PIC, and claim benefits from the commercialization of the genetic resources contained in the natural resources in their lands may well be founded on this latter provision of ILO 169 provision. As already noted, States usually reserve ownership rights over mineral resources and also over genetic resources;[121] in the case of mineral resources, ILO 169 explicitly argues for the right of the indigenous peoples to be consulted, to participate in the benefits, and to receive compensation for any accrued damages.

Despite the low number of ratifications, ILO 169 obviously has the lead in the formulation of international norms for the protection of indigenous

[113] ILO 169 Article 7.1.
[114] ibid.
[115] ibid Articles 5(b) and 8.1; see also ILO 169 Article 17.1.
[116] ibid Article 13.1; ILO (169): Manual (n 111) 29 f.
[117] ibid Article 14.1; ILO (169): Manual (n 111) 31 f.
[118] ibid Article 14.2.
[119] ibid Article 15.1; ILO (169): Manual (n 111) 35 f.
[120] ibid Article 15.2; ILO (169): Manual (n 111) 39.
[121] See Andean Pact Decision 391, Article 6 (see Chapter 5 Section 3).

peoples and its provisions have even induced policy and law-making in countries that have not ratified it.[122] Three major limitations of ILO 169 should be mentioned here. First of all, ILO 169 has implementation problems: most complaints made on the basis of the representation procedure, as stipulated in the ILO Constitution,[123] were associated to lack of consultations, in cases where governments appropriated lands occupied or possessed by indigenous peoples and subsequently granted oil exploitation and/or mining licenses; or to lack of consultations again, when 'development' projects were designed and implemented, which affected the areas inhabited by indigenous peoples.[124]

Secondly, although ILO 169 acknowledges the indigenous and tribal communities' property rights over their lands, it lacks explicit protection of indigenous peoples' intellectual property rights. It acknowledges the 'distinctive contributions of indigenous and tribal peoples to the cultural diversity and social and ecological harmony of humankind',[125] and it supports the promotion of the 'full realization of the social, economic and cultural rights of these peoples with respect for their social and cultural identity, their customs and traditions and their institutions'.[126] Further, the cultures of indigenous peoples are to be safeguarded with appropriate measures,[127] with special focus on the 'social, cultural, religious and spiritual values and practices'.[128] A little more specific is the recognition of handicrafts and other traditional community-based forms of indigenous peoples' activities for the maintenance of their cultures and the call of the Convention for the strengthening and the promotion of such activities.[129] Clearly, the importance of precise intellectual property rights was not as evident at the time the ILO 169 was drafted at present, and the general recognition of their cultures and need for protection was adequate in terms of this international treaty.

Thirdly, ILO 169 lacks explicit reference to the right of self-determination. It seems that this topic was consciously left by the drafters for a wider discussion in the context of the UN.[130]

122 See eg the Philippines Indigenous Peoples Rights Act (IPRA), Chapter 5 Section II.A.2).
123 ILO Constitution Articles 22–24.
124 Swepston in Castellino and Walsh (n 85) 58.
125 ILO 169 Preamble.
126 ibid Article 2.2(b).
127 ibid Article 4.
128 ibid Article 5(a).
129 ibid Article 23.
130 Swepston in Castellino and Walsh (n 85) 57.

### 2. *Other UN bodies and legal instruments on the indigenous peoples' rights*

In addition to the ILO Conventions, general human rights law is applicable with regard to indigenous peoples' rights protection. The human rights activity and gathered experience of the ILO on indigenous peoples' issues is a valuable reference in the more recent UN human rights agenda. Within the UN, there are numerous bodies dealing with the issue.[131] Among the UN umbrella agencies, the World Health Organization has recently become involved in the area aiming at improving the health of indigenous peoples,[132] and UNESCO has an expansive agenda on cultural and educational rights, including those of indigenous peoples, [133] while WIPO has been exceptionally committed to indigenous knowledge questions, folklore and traditional knowledge with the establishment of the Intergovernmental Committee.[134]

Moreover, the UN Special Rapporteur on the situation of human rights and fundamental freedoms of indigenous people was appointed in 2001,[135] with the mandate to gather and receive information and communications from all relevant sources, including governments, indigenous people themselves and their communities and organizations, on violations of their human rights and fundamental freedoms. Finally, the Sub-Commission on the Promotion and Protection of Human Rights appointed Special Rapporteur Erica-Irene Daes in 2002 to specifically examine the issue of indigenous peoples' rights over their lands and natural resources in the inquiry of on the permanent sovereignty of indigenous peoples over natural resources.[136]

#### a) Universal Declaration and the two Covenants

The provisions of the Universal Declaration of Human Rights[137] asserting a general prohibition of any sort of discrimination,[138] the right to property

[131] See the main UN system bodies and documents on indigenous peoples: Indigenous Peoples, The UN System at Work at: http://www.un.org/issues/m-indig.html (26/12/2008).

[132] SeeWHO, Health Services, Indigenous at: http://www.who.int/topics/health_services_indigenous/en/ (26/12/1008).

[133] See UNESCO Portal on Culture at: http://portal.unesco.org/culture/en/ev.php-URL_ID=34603&URL_DO=DO_TOPIC&URL_SECTION=201.html (26/12/2008).

[134] See Chapter 4 Section VI.

[135] Human Rights and Indigenous Issues, Commission on Human Rights Res. 2001/57 of 24 April 2001.

[136] Indigenous Peoples' Permanent Sovereignty over Natural Resources, Sub-Commission on Human Rights Res. 2002/15 of 14 August 2002.

[137] Universal Declaration of Human Rights GA Res. 217 A (III) of 10 December 1948.

[138] UNDHR Article 7: '.... All are entitled to equal protection against any discrimination in violation of this Declaration and against any incitement to such discrimination.'

alone as well as in association with others,[139] and cultural cooperation and development;[140] as well as the provisions of the two International Covenants of 1966, namely the International Covenant on Civil and Political Rights (ICCPR 1966)[141] and the International Covenant on Economic, Social and Cultural Rights (ICESCR 1966)[142] affirming the right to self-determination and protection of minorities; all are regarded by advocates of indigenous people's rights as the basic sources in support of the rights of indigenous communities.[143] The two Covenants are here examined a little more because of their strong connection with the tensions surrounding the right of self-determination.

The ICCPR and the ICESC, in their common Article 1, afford protection to the right of 'all peoples' to self-determination. The provision comprises three paragraphs that relate to the various dimensions of self-determination.

Paragraph 1 of Article 1[144] proclaims the right of self-determination and its main dimensions, namely all peoples' right 'to freely determine their political status' (*political dimension*) and to pursue their 'economic, social and cultural development' (*resource dimension*). The political dimension, in turn, includes an external aspect of sovereignty and an internal aspect of governance[145] that can be linked to Article 25 which requires *democratic* governance; this link has been emphasized by the Human Rights Committee (HRC), the monitoring body established under the ICCPR, in its observations.[146]

Paragraph 2 of Article 1[147] further elaborates the resource dimension of

139 ibid Article 17: '(1) Everyone has the right to own property alone as well as in association with others. (2) No one shall be arbitrarily deprived of his property.'

140 ibid Articles 33, 55, 57, and 73.

141 International Covenant on Civil and Political Rights, GA Res. 2200A (XXI) of 16 December 1966; Scheinin in Castellino and Walsch (eds) (n 85) 3 f; Thornberry (n 85) 116 f and 151 f.

142 International Covenant on Economic, Social and Cultural Rights, GA Res. 2200A (XXI) of 16 December 1966 Thornberry (n 85) 182 f.

143 Other UN instruments to be considered in relation to indigenous peoples in general are the International Convention on the Elimination of all Forms of Racial Discrimination (UN GA Res. 1904 (XVIII) of 20 November 1963), the Convention on the Rights of the Child (UN GA Res 44/25 of 20 November 1989) and the UN Declaration on the Rights of Persons belonging to National or Ethnic, Religious and Linguistic Minorities, GA Res 47/135 of 18 December 1992.

144 ICCPR and ICESC common article 1(1): 'All peoples have the right of self-determination. By virtue of that right they freely determine their political status and freely pursue their economic, social and cultural development.'

145 See also SJ Anaya, Self-Determination as a Collective Human Right under Contemporary International Law' in Aikio and Scheinin (eds) (n 85) 3 f.

146 Scheinin in Castellino and Walsh (eds) (n 85) 9.

147 ICCPR and ICESC common article 1(2): 'All peoples may, for their own ends, freely dispose of their natural wealth and resources without prejudice to any obligations arising out of international economic co-operation, based upon the principle of mutual benefit, and international law. In no case may a people be deprived of its own means of subsistence.'

self-determination through proclaiming the right of all peoples to dispose of their natural wealth and resources. This clause and in particular its last sentence ('...In no case may a people be deprived of its own means of subsistence') has been relied upon by many groups that proclaim themselves as distinctive indigenous peoples in countries where other ethnic groups, especially of European descent are in a dominant position.[148] Further, under the same resources dimension, ICCPR Article 47, as does ICESC Article 25 read that 'nothing in the present Covenant shall be interpreted as impairing the inherent right of all peoples to enjoy and utilize fully and freely their natural wealth and resources'.

Paragraph 3[149] relates to the *solidarity dimension* of self-determination, that is the collective responsibility of State parties to promote the realization of self-determination elsewhere, other than within their own territories.

What is critical in the context of the discourse here for access to genetic resources, when these are located in indigenous lands, and fair and equitable benefit-sharing is the *resource dimension* of self-determination and its association to the *political dimension* of self-determination. However, the problem with the Covenants' text is that the beneficiaries of the right to self-determination are not clear; common Article 1 of the Covenants refers to 'peoples' and not 'States', and there is no definition for 'peoples' neither in Article 1 nor elsewhere, leaving room for different interpretations. Furthermore, the word 'indigenousness' is not included in the human rights texts, and it is thus not obvious whether indigenous peoples should be treated as 'peoples' under Article 1 or as 'minorities' under Article 27.[150]

The use of the word 'peoples' instead of 'States' in the two Covenants created a whole *problematic*: if 'peoples' are defined in a way encompassing indigenous peoples the views vary considerably on what should be the content and the extent of such a collective right, ranging from local autonomy to political independence. Governments have expressed the fear that indigenous communities would be recognized as having right to secession by virtue of this provision and opt for its narrow interpretation.[151]

The *travaux préparatoires* to the ICCPR clarify that the notion of self-determination is framed with colonial peoples in mind.[152] What is more,

148 Scheinin in Castellino and Walsh (n 85) 10.

149 ICCPR and ICESC common article 1(3): 'The States Parties to the present Covenant, including those having responsibility for the administration of Non-Self-Governing and Trust Territories, shall promote the realization of the right of self-determination, and shall respect that right, in conformity with the provisions of the Charter of the United Nations.'

150 ICCPR Article 27: 'In those States in which ethnic, religious or linguistic minorities exist, persons belonging to such minorities shall not be denied the right, in community with the other members of their group, to enjoy their own culture, to profess and practice their own religion, or to use their own language'; see Scheinin in Castellino and Walsh (n 85) 4 f.

151 Swepston in Castellino and Walsh (n 85) 59.

152 Castellino in Castellino and Walsh (n 85) 110.

the collective character of the right to self-determination creates a tricky procedural position:[153] whereas all ICCPR Article 1 cases fall under the mandatory reporting obligation of all States[154] and the inter-State complaint procedure,[155] the collective nature of the right to self-determination prevents the latter from the right of the *individuals* to communicate the violation to the Committee, the reason being the narrow definition of the 'victim' in Article 1 of the Optional Protocol.[156]

However, a liberal analysis of the increasing jurisprudence of the HRC on the issue of the applicability of ICCPR Article 1.2 endorses the argument that the right to self-determination, including eventually the option of determination of the political status, should be made available to indigenous peoples just as to colonial peoples.[157] It is obvious from the recent HRC jurisprudence that ICCPR Article 1 may affect the interpretation of other provisions of the Covenant, including the right of members of a minority to enjoy their own culture found in Article 27.[158] Moreover, the HRC, in its concluding observations on countries' reports,[159] has observed that at least certain indigenous groups qualify as 'peoples' under common Article 1 and favored the right to self-determination of indigenous peoples in the context of 'the means of subsistence' of ICCPR Article 1(2).[160] It

153 Scheinin in Castellino and Walsh (n 85) 12; M Scheinin, 'The Right to Self-Determination under the Covenant on Civil and Political Rights' in Aikio and Scheinin (n 85) 186 f.

154 ICCPR Article 40.

155 ibid Article 41.

156 Optional Protocol to the International Covenant on Civil and Political Rights, adopted and opened for signature, ratification, and accession by General Assembly resolution 2200A (XXI) of 16 December 1966 entered into force on 23 March 1976, in accordance with Article 9. Article 1 reads: 'A State Party to the Covenant that becomes a Party to the present Protocol recognizes the competence of the Committee to receive and consider communications from individuals subject to its jurisdiction who claim to be victims of a violation by that State Party of any of the rights set forth in the Covenant. No communication shall be received by the Committee if it concerns a State Party to the Covenant which is not a Party to the present Protocol.'

157 Scheinin in Castellino and Walsh (n 85) 13 f.

158 See *Ilmari Länsman et al v Finland* ('*Länsman No. 1*') (Communication 511/1992), Views adopted 26 October 1994, Report of the Human Rights Committtee, Vol. II, GAOR, Fiftieth Session, Suppl. No. 40 (UN Doc. A/50/40), 66-76; *Apirana Mahuika et al v New Zealand* (*Mahuica*) (Communication No. 547/1993), Views adopted 27 October 2000, Report of the Human Rights Committee, Vol II (UN Doc A/56/40) 11-29 (particularly paras 9.4 to 9.9); *JGA Diedergaardt et al v Namibia*, (Communication No 760/1997), Views adopted 25 July 2000, Report of the Human Rights Committee, Vol II, GAOR, Fifty-fifth Session, Suppl. No 40 (UN Doc A/55/40) 140–160 (para 10.3); *Marie-Hélène Gillot et al v France* (Communication No 932/2000), Views adopted 15 July 2002, Report of the Human Rights Committee, Vol II, GAOR, Fifty-seventh Session, Suppl No 40 (UN Doc A/57/40) 270–293.

159 Scheinin in Castellino and Walsh (n 85) 11 f to Canada, Mexico, Norway, Australia, Denmark and Sweden.

160 See eg Concluding Observations on Canada, UN Doc CCPR/C/79/Add.105 (1999), para 8: In the HRC concluding remarks to Canada it notes that 'the right to self-determination requires, inter alia, that *all peoples* must be able *to freely dispose of their natural wealth and*

seems that the HRC in interpreting 'peoples' extends particular emphasis in the resource dimension of the right to self-determination of Article 1(2), but not under Article 1(1); clearly, this reserved interpretation intends to avoid any political instability caused by indigenous peoples who live among other groups and claim political self-determination by virtue of Article 1(1).[161]

b} The UN Declaration on the Rights of Indigenous Peoples[162]

The adoption of the UN Declaration on Indigenous Peoples has gone through a very long process: the idea started in the 1980s, the draft text was finished by the Working Group on Indigenous Peoples in 1993, and it was eventually adopted by the General Assembly in 2007. As a declaration, it is an inspirational document for States and the indigenous peoples of the world.

The Declaration implicitly incorporates the substance of the principle of indigenous peoples sovereignty over their lands and resources. Articles 25 and 26 of the Declaration confirm the relationship of indigenous peoples to the lands and waters they traditionally owned or alternatively occupied or used. Further, these provisions recognize the rights of indigenous peoples to ownership, control, and use of the lands, waters, etc. and also the right to full recognition of 'their laws, traditions and customs, land tenure systems and institutions for the development and management of resources, and the right to effective measures by states to prevent any interference with, alienation of, or encroachment upon these rights'.

Article 24 is particularly relevant for the present study because it identifies the indigenous peoples' right to 'their traditional medicines and health practices, including the right to the protection of vital medicinal plants, animals and minerals'; whereas their right to special measures that 'control, develop and protect their sciences, technologies and cultural manifestations, including human and other genetic resources, seeds, medicines, knowledge of the properties of fauna and flora, oral traditions, literatures, designs and visual and performing arts' is provided for in Article 29.

Although the permanent sovereignty of indigenous peoples over their natural resources can be easily implied from the textual composition of Articles 25 and 26, it would have been preferred that the drafters of the text had considered explicit reference to the right, and if they had additionally clarified the issue of subsurface resources.

*resources* and that they may not be deprived of their own means of subsistence' (emphasis added).

161 See critique from Castellino in Castellino and Walsh (n 85) 110; Scheinin in Aikio and Scheinin (n 85) 198.

162 GA Res 61/295 of 13 September 2007.

c) The reports of the Special Rapporteur on Indigenous Rights to Lands

The final report of the Special Rapporteur Erica-Irene Daes, issued on 13 July 2004, for the 56th session of the UN Sub-Commission on Human Rights (26 July-13 Aug. 2004)[163] is pinpointing the developments in the field of indigenous peoples sovereignty over genetic resources. The Report analyzes all the existing domestic and international legal instruments (human rights instruments in particular) and jurisprudence[164] and through this analysis it studies the evolution of the initial political claim for permanent sovereignty to a legal principle and a community right through the two Covenants.

The conclusions of the report are rather revolutionary, because it is actually the first time that indigenous peoples' permanent sovereignty over natural resources is explicitly recognized by a UN policy document, as a 'collective right by virtue of which States are obligated to respect, protect, and promote the governmental and property interests of indigenous peoples as collectivities in their natural resources'.[165] The report highlights the importance of the principle for the future well-being, alleviation of poverty, and social and economic development of the indigenous peoples.[166]

### 3. *Development agencies*

The recognition of indigenous peoples' rights is increasingly promoted by development agencies through their operational policies, such as the World Bank Operational Policy 4.10, accompanied by Bank Procedure 4.10 (OP/BP 4.10),[167] Asian Development Bank Policy on Indigenous Peoples,[168]

163 See Doc E/CN.4/Sub.2/2004/30 (13 July 2004); and Economic and Social Council, Commission on Human Rights—Sub-commission on the Promotion and Protection of Human Rights, 56th Session, Final *Report of the Special Rapporteur, Erica-Irene A Daes*, Prevention of discrimination—Prevention of discrimination and protection of indigenous peoples, Indigenous peoples permanent sovereignty over natural resources, Doc. E/CN.4/Sub.2/2004/30/Add.1, 12 July 2004. The final report should be read in conjunction with Economic and Social Council, Commission on Human Rights, Sub-Commission on the Promotion and Protection of Human Rights, Fifty-third session, Special Rapporteur's final working paper on indigenous peoples and their relationship to land, Prevention of discrimination—'Prevention of discrimination and protection of indigenous peoples and Minorities, Indigenous peoples and their relationship to land', Final Working Paper prepared by the Special Rapporteur, Mrs Erica-Irene A Daes, Doc E/CN.4/Sub.2/2001/21, 11 June 2001.

164 See the Addendum to the Report E/CN.4/Sub.2/2004/30/Add.1.

165 Doc E/CN.4/Sub.2/2004/30, para 56.

166 ibid para 57.

167 See the World Bank operational manual on indigenous peoples at: http://web.worldbank.org/WBSITE/EXTERNAL/PROJECTS/EXTPOLICIES/EXTSAFEPOL/0,,contentMDK:20543990~menuPK:1286666~pagePK:64168445~piPK:64168309~theSitePK:584435,00.html (26/10/08).

168 The ADB policy in available on the Internet at: http://www.adb.org/Documents/Policies/Indigenous_Peoples/default.asp (26/12/08).

and the Inter-American Development Bank Operational Policy on Indigenous Peoples.[169] Accordingly, projects financed by development banks should be designed so as to avoid adverse impact on indigenous peoples, or at least mitigate any adverse effects upon them. There is an effort to essentially involve indigenous peoples in the projects design and implementation. Also, the existing accountability mechanisms have the objective of ex post correcting, if possible, any damage caused and draw lessons for the future.[170]

The United Nations Development Programme (UNDP) adopted specific policies and instructions for dealing with indigenous peoples in their activities,[171] but there is not sufficient information on its implementation.

Non-State actors to date have gained increased importance in the transnational legal process. Given the fact that these institutions play a seminal role in development and that they finance large scale development projects, their instruments despite their internal operational character, have the potential to profoundly influence the way international aid is attributed and conditioned because their policies in practical terms are weighed even more than international law norms.[172]

### *D. Indigenous Rights and Domestic Legislation*

As seen in the previous sections, the overall picture of international bodies and international law related to indigenous peoples lends support to the rights of indigenous peoples in access and benefit-sharing regulation. Modern international law gives useful insights for recognizing indigenous peoples' right to self-determination and to permanent sovereignty over their lands and natural resources, a right deriving from the rights to property and to development.[173]

International norms and the work undertaken by international bodies for the protection of indigenous peoples' rights boost to a certain extent the national process for indigenous peoples' policies; but they cannot stipulate

[169] See the policy at: http://idbdocs.iadb.org/wsdocs/getdocument.aspx?docnum=865801 (26/12/08).

[170] Among others, see eg www.inspectionpanel.org (26/12/08).

[171] UNDP, UNDP and Indigenous Peoples: a Policy of Engagement, 2001 and UNDP—Indigenous Peoples, at: http://www.undp.org/partners/cso/indigenous.shtml (26/12/2008).

[172] GA Sarfaty, 'The World Bank and the Internalization of Indigenous Rights Norms' (2005) 104 Yale Law Journal 1791; L Boisson de Chazournes, 'Policy Guidance and Compliance: The World Bank Operational Standards' in D Shelton (ed) *Commitment and Compliance: The Role of Non-Binding Norms in the International Legal System* (Oxford University Press, Oxford, New York, 2000) 281; B Kingsbury, 'Operational Policies of International Institutions as Part of the Law-Making Process: The World Bank and Indigenous Peoples, in GS Goodwin-Gill and S Talmon (eds) *The Reality of International Law: Essays in Honour of Ian Brownlie* (Clarendon Press, Oxford, New York, 1999) 323.

[173] See also Declaration on the Right to Development, GA Res. 41/128 of 4 December 1986.

detailed provisions to interfere with the relations between the State and the indigenous peoples within its territory. Human rights respect ultimately depends on State policies. Domestic legislative and jurisprudential developments portray the trend for recognition of the indigenous peoples' sovereign rights over their lands and resources better than the international law.[174]

Characteristic is the Philippines Indigenous Peoples Act of 1997, which expressly endorses indigenous self-governance and self-determination within the state.[175] Also, the 1991 Colombian Constitution, together with a series of Constitutional Court decisions, provides for a high degree of autonomy for indigenous peoples and rights related to land, resources, consultation, representation, language and education.

Legislation, however, cannot warrant implementation and enforcement.[176] In the Philippines, the IPRA has been criticized for its controversial provisions and hundreds of human rights violations have been documented even after it was enacted.[177] The Canadian indigenous peoples' policy allows aboriginal groups to benefit from natural resources development projects on their traditional lands with a legal structure that opts for negotiated agreements with a view to fostering indigenous participation and compensation from natural resource development projects on their traditional lands.[178]

In the US, the doctrine of tribal sovereignty dates back to the 19th century with the cases *Cherokee Nation v Georgia* and *Worcester v Georgia*,[179] and recognizes the indigenous communities as political entities.[180] With regards to resources in particular, decisions of human rights jurisdictional bodies, such as the African Commission on Human and

[174] See the Report E/CN.4/Sub.2/2004/30/Add.1.

[175] See also Chapter 5, Section II.A.2.

[176] C Binder, 'Nicaragua: the Case of the Atlantic Coast of Nicaragua: The Awas Tingni Case, or Realizing that a Good Legal System of Protection of Land Rights is No Guarantee of Effective Implementation' in Castellino and Walsh (n 85) 249 f; J Gilbert, 'The Blur of a Distinction: Adivasis Experience with Land rights, Self-rule and Autonomy', in Castellino and Walsh (n 85) 269 f and A Saikia, *Indigenous Control and Sustainability of Common Resources in the Hills of North East India* (Gauhati University, Assam, 2005).

[177] Economic and Social Council, Commission on human Rights, 59th Session, *Indigenous Issues, Human rights and indigenous issues*, Report of the Special Rapporteur on the situation of human rights and fundamental freedoms of indigenous people, Mr. Rodolfo Stavenhagen, submitted in accordance with Commission on Human Rights resolution 2002/6, Addendum Mission to the Philippines, Doc E/CN.4/2003/90/Add.3, 5 March 2003).

[178] Matiation Stefan, 'Growing Teeth: International Law and Indigenous Sustainable Development—The Case of Negotiated Agreements' (September 3, 2004), published on the Internet at: http://ssrn.com/abstract=745784.

[179] *Cherokee Nation v Georgia*, 30 US (5 Pet.) 1 (1831) and *Worcester v Georgia*, 31 U.S. (6 Pet.) 515 (1832), see Report E/CN.4/Sub.2/2004/30 paras 21 f.

[180] See Report E/CN.4/Sub.2/2004/30 para 21: 'Indian nations are 'distinct, independent and political and are qualified to exercise powers not by virtue of delegation of powers from the State government, but by reason of their original tribal sovereignty.'

Peoples' Rights in the *Ogoni* case (involving oil exploration)[181] and the Inter-American Court of Human Rights in the cases of the *Awas Tingni* (involving logging)[182] and the *Maya Indigenous Communities* (involving logging and oil concessions) are illustrative of the recognition of indigenous rights over resources.[183]

The emerging domestic legislation and jurisprudence in tandem with the developing international human rights law on the recognition of indigenous peoples' rights reflects a broader trend for the recognition of indigenous peoples' rights over all resources pertained within the lands they manage. However, genetic, like mineral resources, usually belong by law to the State, and this policy is difficult to change in the near future, although there are some signals towards change.[184]

[181] Communication 155/96, The Social and Economic Rights Action Center and the Center for Economic and Social Rights/Nigeria, paras 55–58 (2001) (Ogoni case), see Report E/CN.4/Sub.2/2004/30 para 49, E/CN.4/Sub.2/2004/30/Add.1 para 18: 'the African Commission on Human and Peoples' Rights found that the Government of Nigeria had violated the collective human rights of the Ogoni people through activities associated with the development of oil resources belonging to the State... Although the State admittedly had the right to produce oil, the 'destructive' and 'repressive' actions of the military government along with the lack of material benefits to the local Ogoni population constituted a violation of (among others) article 21 of the African Charter on Human and Peoples' Rights.'

[182] See Judgment of the Inter-American Court of Human Rights in the Case of the Mayagna (Sumo) Community of Awas Tingni v the Republic of Nicaragua, IACHR, Series C, No 79 (31 August 2001); the Inter-American Court of Human Rights, clearly recognized the indigenous ownership rights to the resources. The Court did not use the term 'sovereignty'. There is no question that the decision found that international law protects the governmental or collective right of the community to the land and resources. The following portion of the judgment is revealing (para. 159): 'Given the characteristics of the instant case, some specifications are required on the concept of property in indigenous communities. Among indigenous peoples there is a communitarian tradition regarding a communal form of collective property of the land, in the sense that ownership of the land is not centered on an individual but rather on the group and its community. Indigenous groups, by the fact of their very existence, have the right to live freely in their own territory; the close ties of indigenous people with the land must be recognized and understood as the fundamental basis of their cultures, their spiritual life, their integrity, and their economic survival. For indigenous communities, relations to the land are not merely a matter of possession and production but a material and spiritual element which they must fully enjoy, even to preserve their cultural legacy and transmit it to future generations.'

[183] *Maya Indigenous Communities of the Toledo District v Belize*, Case 12,053, Preliminary Report No 96/03 (24 October 2003), see Report E/CN.4/Sub.2/2004/30 para 42: 'also was a case aimed at preventing the State's granting of concessions for timber (among other resources) on the Maya lands in southern Belize. The Commission found that the State was obligated by the American Declaration of the Rights and Duties of Man to refrain from granting timber and mineral concessions on land until the Maya lands were recognized and demarcated. ILO Convention No. 169, article 15, provides for indigenous rights to surface resources although to a more limited extent. The land where the resources are located may be land or territory owned by an indigenous people by reason of historic right pre-dating the State, or it may be land acquired in more recent times by purchase, grant, or otherwise.'

[184] See *Awas Tingni v the Republic of Nicaragua* and *Alexkor Limited and Government of South Africa v The Richtersveld Community*.

It was shown above that there is no agreement on the definition of the 'peoples' as subjects of the rights to self-determination, that the HRC makes a clear distinction in the way it construes the *resources* dimension and *political* dimension of self-determination, and that this conceptual debate is largely attributed to the objections of the governments to accept other peoples' sovereign rights within already sovereign States. For the purposes of the current study, the *resources* dimension of self-determination as 'permanent sovereignty over natural resources' is emphasized in support of indigenous peoples' rights to fully participate in the bioprospecting ventures related to the resources they manage, by negotiating and granting PIC and rights to the investors; their participation fleshes out the CBD objectives for fair and equitable benefit-sharing.

But, even if we go beyond the *resources* dimension to a revolutionary recognition of *political* dimension of indigenous peoples' self-determination, it is here argued that accepting in international law that literally all peoples have a right to self-determination does not mean that all are entitled to change the status quo of political ordering. In fact, independence does not seem to be an issue for most indigenous groups.[185] The true right to self-determination is not determined from the outward form of the self-governing or administrative institutions of indigenous peoples, but from the inward set of values of perceived universal applicability, founded on the belief that all humanity, individuals and groups, have the right to pursue their own destinies in freedom and under equal conditions.[186] The test of peoples' self-determination is a more subjective process depending on whether they actually feel that they have choices about their way of life.

The tensions with respect to self-determination in the international arena mark an opportunity as well as a difficulty in the evolution of international law today and call for a review of the notion of 'self-determination' from an end-state to a relational approach.[187] This review would not put in question the norm itself, but would adapt it to modern reasoning. The Declaration includes much of the material for the reconstruction of the concept in relational terms, but this material needs still to be developed. The recognition of permanent sovereignty of indigenous peoples over resources does not contradict the sovereignty principle proclaimed at the CBD:[188] in the CBD, State sovereignty over the resources differs from

[185] Scheinin in Castellino and Walsh (n 85) 10: the Quebec Secession Case confirmed that the right of secession is recognized only under specific conditions; Thornberry in Aikio and Scheinin (n 85) 64.

[186] Anaya in Aikio and Scheinin (n 85) 8; Kingsbury in Aikio and Scheinin (eds) (n 85 2000) 22.

[187] Kingsbury in Aikio and Scheinin (n 85) 19 f.

[188] See different opinion in D Harry and LM Kanehe, 'The BS in Access and Benefit-sharing (ABS): Critical Questions for Indigenous Peoples' in B Burrows (ed) *The Catch: Perspectives on Benefit-sharing* (Edmonds Institute, Edmonds, Washington, 2005) 81: the authors believe

recognition by States of the permanent sovereignty of indigenous peoples over the resources they manage. Recognition of indigenous rights over the lands and resources does not deny the State's authority to determine access to genetic resources within its borders and to enact legislation on access and benefit-sharing. The State remains the sovereign authority supervising bioprospecting projects undertaken within its territory and involving indigenous communities. In the capacity of supervisor, the State should protect the rights of its citizens, so as to ensure their participation in the process, and warrant that benefits derived from the use of extracted resources reach those responsible for their conservation, management and sustainable use.

## IV. PROCESS AND BENEFITS DISTRIBUTION IN ACCESS AND BENEFIT-SHARING AGREEMENTS

### *A. Process*

#### *1. Factors in the negotiation of the bioprospecting agreements*

Contract law structure offers useful insights on the access and benefit-sharing process: a number of factors may influence the parties' behavior and may raise difficulties because of the parties' uneven bargaining position. Since the bioprospecting venture presents a characteristic example of the uneven situation between the private sector and the developing country, and in particular the indigenous peoples, it is useful to look at some basic contract law features.

One of the complexities in designing a contract is its zero sum nature: what one party gains, the other loses. This means that the contract terms must be structured carefully to apportion risk according to relative risk-aversion, while providing some assurances against adverse selection and moral hazard,[189] where risk is understood as potential harm that may be caused by the flow of the events. While 'risk' is often understood as 'probability', 'risk' has a more negative nuance than 'probability'. Risk aversion is a concept in economics, finance, and psychology that explains the behavior of consumers and investors under uncertainty; it is the reluctance of a person to accept a bargain with an uncertain payoff rather than another bargain with more certainty but possibly lower expected payoff. The uncer-

that CBD sovereignty principle and its recent reaffirmation in the preambular language of the COP VII decision on an international regime on access and benefit-sharing (Decision VII/19) sets an unsafe stage for future negotiations of an ABS regime.

[189] This section draws particularly on W Lesser, *Genetic Resources—Convention on Biological Diversity: Exploring Access and Benefit-Sharing Issues* (CAB International, Oxon, New York, 1998) 36 f.

tainty in terms of probability and time of producing a valuable commercial product is very high in the pharmaceutical industry. The supplier and the purchaser or user of genetic material invest in the hope of discovering another blockbuster drug, but for both there is a high probability-risk of losing their investment.

Parties' attitudes to risk vary, from risk averse, to risk neutral, to risk-loving or risk tolerant. For instance, there is choice between either (i) receiving 1,000 or nothing, with a probability of 50 per cent; or (ii) 500 with a probability of 100 per cent. The person that chooses the first option is risk loving or risk tolerant, whereas the person that chooses the second option is risk averse. Finally, risk neutral is the person who is not particularly concerned with either option (i) or (ii).

With regard to bioprospecting for medicinal products, companies are usually risk neutral, meaning that they have a symmetrical response to earning or losing a sum of money, whereas source countries and communities are more risk averse, as they respond much more strongly to the chance of losing a sum than to earning a like amount. Besides, the fact that neutral parties should bear most risk for purposes of efficiency can be illustrated by mathematics. A family with a $4,000 budget would not be concerned for the loss of a $100 bill, but a family with an $800 budget would. Large companies intend to be well-financed and moderate losses are not devastating. Also, their business model is run on the basis of risk assessment and mitigation, and a number of projects are implemented simultaneously according to risk management principles.

Apart from considerations relating to risk, information asymmetries influence the equity and structure of the agreements. If one party is not adequately informed, it may select 'bad' products or customers (adverse selection). Finally, there is always a risk that the other party's behaviour is unreliable, in a way that the consequences of such behaviour do not harm it, but may even benefit it (moral hazard). In access and benefit-sharing contracts, the industry has an informational advantage, so the question is how to better inform and educate the source countries and the indigenous peoples to reach an equitable agreement.

### *B. Agreements' Structures*

Depending on the circumstances and the context, the structure and forms of the bioprospecting agreements differ. The main concern though in the design of the agreements is the identification of the owner and correspondingly of the beneficiary. This is however still a very difficult task, because of unclear land tenancy systems and/or non-recognition of customary land tenure systems.

Of central importance in the access and benefit-sharing process is the granting of the PIC. Under the CBD, PIC applies in two distinct cases: that of access to genetic resources[190] and that of bio-safety.[191] We saw that the CBD hardly explains the process and the conditions for the PIC. Another MEA where PIC is required is the Basel Convention.[192] Under the latter, the requirement for PIC is far more explicit than the CBD:[193] each signatory must identify a national authority for the procedures administration[194] and the information to be provided is further specified.[195] The implementation of PIC under the Basel Convention can hardly be used as guidance for the implementation of this requirement under the CBD. The legal design of the access and benefit-sharing undertaking may vary from one sole contract to a complex of different contracts. Actually, contracts can be negotiated on a case-by-case basis because there are a lot of particularities not to be covered by a single model. This is an additional reason for the impossibility of a 'one model' contract prescribed by an international treaty/protocol to cover all different cases.

MTAs are already a standard tool in commercial and academic research partnerships. The MTA approach is centered on the transfer of materials, begins with the source country property, and lends itself to greater control of the genetic materials by the source country.[196] Another type is the Research/Development agreements, broadly used in the ICBG practice, where the focus is placed on the research process, rather than transfer of materials.

The ICBG is illustrative of these difficulties;[197] although the policy objectives of all the ICBG projects are the same, there is a great diversity of contracts according to the stakeholders, the subject matter and the circumstances. The 'one contract' model, with the Costa Rica ICBG being a very good example,[198] is highly complicated and demands technical sophistication and multi-disciplinary knowledge. Its operation was eventually possible thanks to the broad mandate to INBio from the Costa Rican Government and the general built up experience of the institute.[199]

190 CBD Article 15.5.

191 ibid 19.3.

192 Basel Convention on Transboundary Movements of Hazardous Wastes and Their Disposal.

193 Lesser (n 189) 86 f; on the Informed Consent in Medicine and the three accepted theoretical foundations, see ibid 88.

194 See Basel Convention Articles 6 and 7.

195 ibid Annex V A.

196 DM Puttermann, 'Model Material Transfer Agreements for Equitable Biodiversity Prospecting' 7 (1996) Colorado Journal of International Environmental Law and Policy 141.

197 JP Rosenthal (1998) 'Activities of the International Cooperative Biodiversity Groups Program. A Benefit-Sharing Case Study for the Conference of Parties to the Convention on Biological Diversity' Submission to the COP 4 of the CBD.

198 Parties to the agreement were: Cornell University, the Cornell Research Foundation, INBio, and the Bristol-Myers Squibb Pharmaceutical Research Institute.

199 Rosenthal (n 197).

To the other extreme a wheel of bilateral benefit-sharing agreements was chosen for the Latin American ICBG.[200] The choice for such a structure was inevitable because the programme actually involved agreements with three different Latin American countries, Chile, Argentina and Mexico, and no one contract could accommodate terms for each country.

A dual contract model close to the bilateral approach was chosen for the Suriname ICBG contract[201] and the African ICBG.[202] Despite some differences, such as the fact that there is no major commercial partner in the African ICBG, these agreements clearly separate between the Research/Development Agreement and the Collections/Benefit-Sharing Agreement and one or more host country institutions are involved in both agreements, such as the Conservation International in the ICBG Africa and the Bedrijf Geneesmiddelen Voorziening Suriname, a pharmaceutical company owned by the Surinamese Government, in the Suriname ICBG.

The Peru ICBG constitutes a unique example in terms of the history of bioprospecting negotiations; the arrangements ended up to a relationships triangle of sorts, where the collection/benefit-sharing agreement (BCA) was separated from the financial benefits (license option agreement) and the use of the traditional knowledge (know-how license). The question arising here is which model grants the greatest efficacy and warrants the most fair and equitable sharing under the CBD: MTAs, Research Development Agreements, Collections Agreements or the 'one contract model', bilateral contracts or other contractual types linking the parties according to the private will and needs.

On the one hand, bilateral agreements are easier to negotiate, because they can more easily accommodate the diverse interests of the parties than the 'one contract' model. They also grant greater stability because if one partner changes or wishes to change the terms on a certain subject matter, this usually does not affect the other agreements. The problem is that there is a lot of burden placed on the 'hub' institution coordinating the project. On the other hand, the 'one contract' approach like in the Costa Rica ICBG is not applicable in all cases, because of its complexity, which poses problems for the locals to keep control of the situation. Complex 'one contract' bioprospecting arrangements are contingent upon the educational level of source countries and in particular of local communities and their capacity to handle such complex legal situations. The advantage of such a model lies in that a single contract may increase the bargaining power of local groups

200 ibid.

201 M Guérin-McManus et al, 'Bioprospecting in Practice: A Case Study of the Suriname ICBG Project and Benefits Sharing under the Convention on Biological Diversity' (1998) Case Study submitted to the CBD Secretariat.

202 M Iwu and S Laird, 'The International Cooperative Biodiversity Group—Drug Development and Biodiversity Conservation in Africa: Case Study of a Benefit-Sharing Plan', Case study submitted in response to the call for benefit-sharing case studies by the Secretariat of the Convention on Biological Diversity (1998).

over specific commercial terms in licensing and research agreements.[203] In any case, even when more than one agreement is negotiated, it is important that local parties require and have the opportunity to examine the related contracts and be involved in any terms they deem important as a condition of their participation.

## *C. Benefits*

### *1. Monetary benefits*

Monetary benefits include royalties, advance payments, and milestone payments.[204] A percentage of earnings from commercial sales by the licensing partner (royalties) may be agreed upon in the initial agreement, or the agreement may specify a range and require parties to negotiate the final rate on a case by case basis. Some issues to consider in royalty structures include the relative contribution of partners to invention and development, the information provided with samples, and the novelty or rarity of the sample organism.

Evidently, the more the commercialized product is closer to the original extract (direct isolate of the original extract, very similar, or synthesized derivative from the original lead) provided by the source country partners, the higher the royalty payment is. Furthermore, the screening strategy used in the activity identification is another factor of royalty rates definition; the use of a random, broad-based taxonomic, ethnobotanical, of functional, or mechanistic approach each time determines the proprietary rights of the contributors.[205]

Full awareness and consideration of the above factors in the royalty rates definition determine the degree to which royalty schedules are fair. In the ICBG practice, contracts include royalty schedules that specifically provide for the reward of the source country, depending exactly on the relation of the end product to the initial one and on the screening strategy used. The timing of the negotiations on the royalties may affect how fair and equitable the transaction is. Given the uncertainty of drug R&D, the best time for the negotiation of royalty schedules is after the compound has been chemically analyzed and data on its therapeutic activity are available. It is recommended that contracts allow for flexibility in the royalty negotiations at later stages.

[203] See also R Korobkin, 'Bargaining Power as Threat of Impasse', Research Paper No. 04-6, University of California, Los Angeles School of Law, Law and Economics Research Paper Series, February 2004.

[204] See in particular Rosenthal (1998).

[205] About the random, biorational and ethnomedical approaches to drug discovery, see Rosenthal in Grifo and Rosenthal (eds) (1997) 281.

As the commercialization of a product may never happen or may happen after a very long period of time, it is not appropriate to limit monetary benefits in royalty mechanisms alone. Drug R&D, final approval, and market launch might take years and therefore access fees in the form of advance orup-front or lump-sum/milestone payments should be awarded to the source country or local people for the provided genetic material and the associated medicinal knowledge, to secure compensation independent of future success of the process.

Advance payments, being payable irrespective of whether or not the final product succeeds in the marketplace, offer an increased possibility of equitable participation by indigenous peoples in the benefits associated with the use of their resources. Advance payments are valuable for establishing trust funds that can provide immediate benefits to stakeholders. Milestone payments constitute a kind of advance non-returnable royalty payments, contingent upon the potential new product reaching the advanced stages of drug development, eg in the US an Investigational New Drug Application and a New Drug Application for a Licensed Product with the US Food and Drug Administration.

Due to adverse selection, it is difficult to identify an equitable/appropriate distribution between monetary benefits, especially lamp sum and royalty payments. A real solution would involve improved information, but the risk aversion of the supplier of the resources and the risk neutrality of the buyer/user speak for the up-front type as a preferred type of monetary compensation. Up-front payments benefit the providing country and give an incentive for conservation and careful use from the beginning of the contract relationship. Milestone payments stand in between royalty payments and up-front payments providing compensation to the suppliers and recognizing their contribution in material to the progress in R&D.

As discussed below, non-monetary benefits can actually be much more valuable, as they add value to the primary resources in the country of origin. Up-front payments could have the form of training and equipment. However, monetary benefits are still necessary to the extent that countries and communities may urgently be in need of money.

### a) Trust fund mechanisms

Trust funds, the mechanisms that collect advance payments, and any prospective royalties, constitute one of the most flexible and equitable ways to manage monetary benefits in both the near and long-term. Most benefit-sharing schemes today include trust funds, because a certain degree of fairness in the distribution of benefits is secured through projects designed and implemented for the common interest of the community. Trust funds represent communities and national interests and most of them have been able to advance sustainable development projects[206] through the financing of projects on biodiversity management, health and local development.

The objective of a trust fund is to guarantee that the locals actually benefit from the access and benefit-sharing undertakings. However, literature in economics argues that even when locals appear well-positioned to seize the ex post profits of a successful bioprospecting activity, this does not mean that local development will be fueled nor that poverty will be reduced.[207] In fact, it seems to be hard for locals to participate in the new market structures, and this is at least the outcome of the Argan Oil case in Morocco: opening new markets for local products may induce indigenous product differentiation that easily excludes locals, in particular the poor.[208] If locals participate in the new market, they might do so in a superficial manner and the final distribution of benefits does not equally reach regions or households.[209] Beyond these considerations, it is very difficult to define in advance how the ultimate goals of conservation and sustainable use are addressed. This is possible when the losses in biodiversity and natural resources occur at rates within their capacity for renewal.

### 2. *Non-monetary benefits*

Apart from the monetary compensation of the countries contributing in tangible genetic resources and intangible medicinal knowledge, economic growth and development should be primary objectives in the bioprospecting practice. Commercial entities or non-profit organizations may provide resources to help build the capacity of partner entities in the source country to execute current or future needs for bioprospecting research, medical

[206] In the Vietnam-Laos ICBG the money is spent for the health and education primary materials and infrastructure, see Soejarto et al in JBA/NITE (2002) and Soejarto et al (2002); In the African ICBG a local clinic was rebuilt, tools were purchased for preparation, and storage of herbal medicines in a community pharmacy and cultivation of medicinal hedges, see Iwu and Laird (n 202); in the Suriname ICBG, the money was used for tools, training and marketing for traditional non-timber forest craft products, and a meeting of indigenous leaders was organized to foster common interests and finance shaman apprenticeships, see Guérin-McManus et al (n 201); K Ten Kate and A Collis, Benefit-sharing Case Study: The Genetic Resources Recognition Fund of the Univeristy of California, Davis, Submission to the Executive Secretary of the Convention on Biological Diversity by the Royal Botanic Gardens, Kew; P Bailey, Fund to repay developing nations for valuable genes, Dateline UC Davis (7 March 1997); S Rubin and S Fish, 'Biodiversity Prospecting: Using Innovative Contractual Provisions to Foster Ethnobotanical Knowledge, Technology, and Conservation'(1994) 5 Colorado Journal of International Environmental Law and Policy 23.

[207] TJ Lybbert, CB Barrett and H Narjisse, 'Who Benefits from a Bioprospecting-Based Boom? The Case of Argan Oil in Morocco' (January 2001) published on the Internet at: http://ssrn.com/abstract=257326.

[208] ibid.

[209] ibid.

care, and biodiversity management. Non-monetary benefits include training, equipment, and infrastructure development. These efforts are targeted at the improvement of the capacity of the source countries to add value to the genetic resources, ameliorate their input in the medicines production, and gain greater benefits in the future.[210] Ideally, agreements require that locally important, but understudied, diseases and indigenous therapies will be investigated by commercial and other scientific partners.

## V. CONSIDERATIONS ON INTERNATIONAL COOPERATION AND BIODIVERSITY GOVERNANCE

In the same spirit as the Kyoto Protocol, the CBD has embraced the use of market-based economic realities.[211] Given the difficulties for the negotiation of a single international treaty on access and benefit-sharing, domestic legal frameworks are the first step to protect developing countries' interests in the genetic resources market. At the same time, such frameworks can be most effective only with the participation of the industry and research institutions. In any case, states should not unduly restrict bioprospecting undertakings, but should even enhance the opportunity for pure scientific research.

In the meantime, coordination is necessary at the international level, not only to synchronize the action taken within the different fora, but also to keep track of the developments in the access and benefit-sharing activities throughout the world. There are efforts underway to build an international binding instrument on access and benefit-sharing. The already described international activities at the international fora and the many difficulties encountered because of the different actors' opposed interests provide empirical evidence that such an international document is a very ambitious duty at the moment.

There is a fear that if we try to design an international treatyor protocol on access and benefit-sharing as soon as possible, this will lead to interpretation and implementation problems because, when delegates are closed in a room during a summit or other conference with the mandate to come up with a text by the next morning, they will bargain and negotiate a win-win solution that appears to accommodate all States' concerns, in order to meet the deadline.

[210] See in particular Rosenthal (n 197).

[211] Also MI Jeffery, 'Special Feature–Ten Years after Rio: Implementing Sustainable Development–Bioprospecting: Access to Genetic Resources and Benefit-sharing under the Convention to Biodiversity and the Bonn Guidelines' (2002) 6 Singapore Journal of International and Comparative Law 747.

Furthermore, legal and political analysts very often forget the fact that international players have the tendency to change the forum of negotiation, when their interests are not fully met. This forum and regime changing leading to regime complexity is a feature of international intellectual property law and currently also access and benefit-sharing.[212]

Even the Bonn Guidelines, the soft law rules on access and benefit-sharing, have been criticized for failing to protect the rights of all stakeholders, by not representing the majority of indigenous peoples: at the COP VI, most indigenous peoples decided to abstain from the discussions on the Bonn Guidelines and reject their implementation because, in their view, participation in the Guidelines' development meant facilitation of biopiracy of their own resources and knowledge.[213] Apart from these observations, international relations and political economy discourse, together with some other examples of the legalization of international governance in other issues, lend support to the argument that at the present time a binding instrument on access and benefit-sharing may not be a realistic option.

Currently, the fundamental acknowledgement in international law and politics of the interdependence of the States and the need for collaboration to enhance global welfare and mitigate substantial global problems that need international action, such as environmental problems, have led to an impressive degree of legalization of international relations. By entering into 'covenants', by delegating authority to international institutions for the interpretation and the implementation of the law, and by relinquishing part of their sovereignty, States manifest their normative commitments to international cooperation and action.

The value of biodiversity as a global common good and the need for international binding regulation on access and benefit-sharing are rightly contemplated. To be pragmatic, considering the reluctance of global players, it is evident that since optimal collective action is not possible, suboptimal unilateral action by larger sub-global governments is better than no action at all.[214]

International norms vary from *hard law* instruments, which refer to binding and precise obligations[215] to purely political arrangements in which

[212] K Raustiala and DG Victor, 'The Regime Complex for Plant Genetic Resources' International Organization, Spring 2004, UCLA School of Law, Research Paper 03-19, published on the Internet at: http://ssrn.com/abstract=441463; LH Helfer, 'Regime Shifting: The TRIPs Agreement and New Dynamics of International Intellectual Property Lawmaking' (2004) 29 Yale Journal of International Law 23.

[213] Harry and Kanehe in Burrows (ed) (n 188).

[214] KH Engel and SR Saleska, 'Subglobal Regulation of the Global Commons: The Case of Climate Change' (2005) 32 Ecology Law Quarterly 183.

[215] On the rationales of hard law see KW Abott and D Snidal, 'Hard and Soft Law in International Governance' (2000) 54 International Organization 424 f, where they use contracting theory to argue against the sharp bifurcation between rationalistic and construc-

legalization is largely absent,[216] whereas in-between there is a broad category of the so-called *soft law* instruments.

The form chosen by the international actors, including States and firms, but also activists' organizations and other non-state groups operating at both domestic and international levels,[217] is defined by the efficacy of the instrument in different circumstances and by the acceptance of certain international legalization of the issue through international arrangements. The wide variety of international legal arrangements is illustrated in international relations with examples characterized along the hard law/soft law continuum in economic terms, with the triplet {O,P,D}, referring to the three dimensions of obligation, precision and delegation; high levels of each dimension are indicated by capital letters {*O,P,D*}, moderate levels by small letters {*o,p,d*}, and low levels by dashes {-,-,-}, thus expressing the variations of each dimension in the triplet.[218] Contracting costs, sovereignty costs and uncertainty and risk factor define the *softness* of the instrument selected by the international community.

Particularly for new and complex issues, such as access and benefit-sharing, soft law has the advantage of low contracting costs.[219] Indeed, the evolution of the ILO, the International Atomic Energy Agency (IAEA) and world trade show the impact of contracting costs.[220] First the ILO, because of the States' reluctance to ratify its convention, emphasized non-legally binding instruments, in the form of codes of conduct and recommendations, with view to reducing national ratification costs and alleviating the organization's prestige. Secondly, in the case of the IAEA, the sensitive issue of protection of nuclear materials and technical matters are treated with recommendations. Last, but not least, in international trade, States resisted to negotiate and finally draft the initially proposed charter for an International Trade Organization at Bretton Woods in 1944 and promoted the adoption the low-cost interim framework of the 1947 GATT for tariff reductions. In the case of international trade, it actually took 47 years for the states to agree on the advantages of harder legalization in this delicate area and to create the WTO in 1994. Access and benefit-sharing is a also a highly sensitive and complicated area and given that the CBD put the foundation only in 1992, there is still a long way to go for an international binding tool.

The softness of the instrument is further defined by the sovereignty costs to be endured.[221] The traditional Westphalian concept of sovereignty

tivist/normative approach: the origins of international legalization are both an interest-based and a normative enterprise.

216 On the advantages of soft law, see ibid 434f.

217 ibid 450f.

218 ibid 421.

219 ibid 434–436.

220 ibid.

221 ibid 436–441.

suggests that the more external authority States accept in important decision making, especially in their dealings with other states and in matters related to the citizens within their territory, the greater the sovereignty costs are. In areas such as human rights and territorial issues, social security,[222] and technical matters,[223] States try to limit sovereignty costs through arrangements that are non-binding or are imprecise or do not delegate extensive powers. Useful examples are the UNCLOS on the re-definition of national territory by innocent passage rights, the Council of Europe framework Convention for the Protection of National Minorities, the international money-laundering regime, and the 1989 OECD Financial Action Task Force of financial experts. An efficient treaty on access and benefit-sharing would include obligations on the treatment of indigenous peoples, and States are not ready to accept such an intervention on their relations with their citizens.

Finally, soft law is an option to deal with uncertainty[224] and risk.[225] The imprecise commitments found in MEAs show how states deal with these variables.[226] In bioprospecting activities, because of the diverse actors, values, and interests, the information asymmetries and the risk pertaining to the outcome of the undertakings, soft law in the international governance can facilitate compromise and beneficial cooperation.

The learning process from soft to hard legalization is also illustrated by the FAO-UNEP regime requiring PIC to international transfers on hazardous chemicals and pesticides. The FAO adopted a Code of Conduct in 1986 on the distribution and use of pesticides, while UNEP adopted 'guidelines' in 1987 on the exchange of information on internationally traded chemicals. In 1989, FAO and UNEP coordinated amendments to their instruments and decided to add and manage jointly PIC requirement and procedure for international transfers of hazardous substances. The suggested PIC system would involve a low obligation, relatively high precision, and moderate/largely administrative and technical delegation {-,*P*,*d*}. Despite extensive consultations with expert groups, governments, and industry; and technical assistance, the attempt actually to 'harden' the

[222] ibid, eg in the most institutionalized alliance ever, NATO, the delegation is moderate and in it the EU security affairs have lagged behind in the integration process.

[223] ibid Technical complexity makes it hard to agree on international transportation and food standards, eg ISO rules or within the EU the EMAS Regulation.

[224] ibid 441–444.

[225] The classical distinction between uncertainty and risk is that when risk is the central concern (ie actors cannot predict the outcome of an agreement, but know the probability distribution of several outcomes, conditional on agreements terms) precise agreements offer a way to manage and optimize risk-sharing. But, when circumstances are fundamentally uncertain (ie when even the range and/or distribution of possible outcomes is unknown), a more precise agreement may not be desirable.

[226] See the Vienna Ozone Convention, the Agenda 21, Forest Principles and the Framework Conventions on Climate Change and the Protection of Biological Diversity.

system failed in Rio in 1992. FAO and UNEP continued to administer the PIC system until member states of both organizations authorized formal negotiations and a convention was approved in September 1998.[227]

Following this discussion, it is argued here that we still find ourselves in a learning process for access and benefit-sharing. Although non-binding, the Bonn Guidelines in conjunction with an international coordinating agency and coordination points within states, where private and other non-governmental actors usually adhere can be successful by making international rules on access and benefit-sharing more precise as learning and experience occur. The CBD Secretariat is already a very important information provider on access and benefit-sharing, with its access and benefit-sharing case studies and domestic legislation databases freely available on the Internet.

On the basis of changes in corporate culture, a policy document entitled the 'Access and Benefit-Sharing Management Tool' (ABS-MT), was conceived to help facilitate the negotiations and implementation of fair and equitable access and benefit-sharing arrangements according to the CBD and the Bonn Guidelines. The draft ABS-MT was developed during 2003/04 by the International Institute for Sustainable Development (IISD) team with input from experts and practitioners. Its elements are a set of substantive requirements to guide ABS practices, a management system to guide its structured application, and an outline of issues and approaches relevant to the assurance process.[228]

A 'Facilitating Mechanism', on the basis of a voluntary sharing expertise on a regional basis, functioning as an 'honest broker' and mediating information on related to bioprospecting legal, institutional, business, scientific and conservation aspects[229] could be an essential regional initiative, in coordination with a global Clearing House Mechanism within the CBD Secretariat.

In a similar context, responding to the mediation and facilitation set out in Article 48 of the Bonn Guidelines, notable were the efforts made by the Swiss government to include a 'mediation mechanism', in the form of a governmental institution or private non-profit entity, aimed at facilitating

[227] Abott and Snidal (2000).

[228] See Doc UNEP/CBD/WG-ABS/3/5; State Secretariat for Economic Affairs and International Institute for Sustainable Development, A Guide to Using the Working Draft ABS Management Tool (SECO/ISIC, Berne, 2005), published on the Internet at: http://www.iisd.org/pdf/2005/standards_abs_mt_user_guide.pdf (23/12/08) and J Cabrera, G Greene and T Rotherham, ABS Management Tool Project: Summary Project Description, *International Expert Workshop on Access to Genetic Resources and Benefit-sharing, IV. Instruments/tools/measures which could assist in achieving the IR: Company conduct, standards and certification*, Cuernavaca, Mexico (24–27 October 2004).

[229] AF Krattiger and WH Lesser, 'The 'Facilitator': Proposing a New Mechanism to Strengthen the Equitable and Sustainable Use of Biodiversity' (1995) 22 Environmental Conservation 211.

negotiations between stakeholders operating in fields open to mutual agreements. The objectives of this mechanism were to promote fair and equitable solutions in each particular case on the issues of access to genetic resources and benefit-sharing, and to contribute generally to development and technology transfer. Participation of the parties would have been on a voluntary basis. In the end, this initiative was not adopted, but it is an option which might more seriously be taken into account in the future.[230]

It has also been suggested that countries rich in biodiversity form a cartel similar to the Organization of Petroleum Exporting Countries (OPEC) on the grounds that many resources can be found in more than one country.[231] This proposal can hardly flourish because the pharmaceutical industry is much less dependent on natural resources than the oil industry, which is forced constantly to bargain with State governments and OPEC on the conditions of access to greatly demanded petroleum reserves, as there is no synthetically engineered product competition to date. For drug companies, bioprospecting is just one among many economic strategies, and not necessarily the preferred one. Moreover, the extremely wide geographical distribution of the species, the reproduction and high portability of genetic resources that makes the tracking of their movements hard, the fact that in bioprospecting it is to a very large extent the information that actually 'makes' the resource useful, rather than its physical content and that groups of countries would be trying to control not one single natural resource with proven value, like oil, but many resources, almost none of which has any proven significant commercial applications, renders such a proposal unrealistic.

## VI. SUMMARY AND CONCLUDING REMARKS

Bioprospecting activities are the basis for the research and development of a large number of biotechnology products today, such as drugs, cosmetics, and chemicals. In the field of drug development, more specifically, there is already a number of drugs produced from biological resources and there is still a great potential for the discovery of plants, animals, micro-organisms and other biological resources with valuable medicinal properties.

Realizing the commercial value of the use of genetic resources, the idea is to provide incentive measures to promote their conservation and sustain-

[230] B Chaytor, R Gerster, and T Herzog, 'The Convention on Biological Diversity – Exploring the Creation of a Mediation Mechanism' (2002) 5 The Journal of World Intellectual Property 157.

[231] See in particular S Biber-Klemm, P Cullet and K Kummer Peiry, 'New Collective Policies' in S Biber-Klemm and Th Cottier *Rights to Plant Genetic Resources and Traditional Knowledge—Basic Issues and Perspectives* (CABI Publishing, Berne, 2006) 283.

able use. Studies in environmental economics underscore the significance of economic instruments, such as market creation, creation of property rights, and better market information through valuation techniques for biodiversity conservation. To achieve the public goals of biodiversity conservation and sustainable use, access and benefit-sharing measures should take advantage of the existing market in genetic resources and use it to channel the benefits to all the contributors for the product development.

Biogenetic resources of medicinal value are primarily collected from developing countries whose local and indigenous populations have developed knowledge associated with these resources. The 'biopiracy' talk, which has become more heated during the last decades, describes the unfair situation in which the benefits from the commercialization of such resources and knowledge do not reach those who manage the resources and hold any knowledge on the properties of such resources. As a result, they are given no sufficient incentives for the conservation and sustainable use of the surrounding ecosystem and opt for more profitable, but destructive, activities.

In order to invest in the designing of policy measures that use the market potential on bioprospecting, a valuation exercise is necessary. In fact, there is a debate going on with regard to the medicinal value of genetic resources, and the use of bioprospecting activities for conservation purposes. The great uncertainty in pharmaceutical R&D, the informational character of genetic resources differentiating them from other biodiversity goods and services, and the vast array of production methods and technologies available in the drug industry are factors that reduce the importance of bioprospecting activities for drug development. Nevertheless, due to the diversity of climatic and geological conditions on the globe, organisms developed biological processes, defenses and substances of importance for future pharmaceutical R&D that can hardly be conceived or imitated in the laboratory. Apart from pharmaceuticals, the market for botanical drugs, which are exclusively based on natural lore, is taking even broader dimensions and gains the preference of consumers, such as Chinese traditional medicine or Indian ayuverda. The chances of discover a drug are diminished if species are extinct.

The fact that genetic resources were seen as free to access and use before 1992 underpins the 'biopiracy' problem. In fact, nothing in international law speaks against the sovereign rights of States over the genetic resources extracted from the biological resources found within their territories. However, free bioprospecting has been practiced for centuries and the 1983 International Undertaking placed genetic resources in the common heritage of mankind. The common heritage of mankind does not contradict sovereignty rights, but it was commonly understood in international practice that genetic resources belong to the commons and are free to access. Later, when technological innovation led to a growing number of biotechnological products based on bioprospecting undertakings, these products were vested with intellectual property rights and no benefits were returned to the source country for its contribution in material and information.

The 1992 CBD came to restore the balance between the users and providers of genetic resources proclaiming State sovereignty over genetic resources while States undertake the international responsibility to guarantee the protection of biodiversity within their boundaries. The CBD sets the objectives of biodiversity conservation, sustainable use, and fair and equitable sharing of the benefits arising from its use. Its access and benefit-sharing provisions conditioning access and benefit-sharing on PIC and MATs especially are of an aspirational character and signatory states are called to enact policy and legislative measures to flesh out the objectives of the Convention.

Due to the broad manner in which the CBD defines its objectives, it was deemed necessary within the current thesis to examine in a multidisciplinary and inter-temporal manner the objectives incorporated in the first article of the CBD in order to shed light on the implementation options that better serve its premises. In particular, the search for the real meaning of the 'fair and equitable' formula developed in international law and adjudication presents many challenges. The CBD is emphatic on its third objective for 'fair' and 'equitable' benefit-sharing matching the 'fair process' and the 'equitable distribution' of the deriving benefits. In implementation terms, this is further translated into guaranteeing democratic processes in the bioprospecting ventures, thus allowing all stakeholders to equally participate in the negotiation of the agreements; and into protecting fundamental human rights, such as property and cultural rights, rights to development, and indigenous rights.

In pursuit of an international access and benefit-sharing instrument, we witness intensive action within the ABS Working Group in coordination with the Working Group on Article 8j instructing the COPs, the WIPO IGC, and the WTO TRIPS Council. However, the diverging interests as reflected in the negotiations of the above mentioned fora, the geographical and geopolitical conditions, the particularities of the different industries (medicines, chemicals, cosmetics, agriculture, horticulture) active in bioprospecting and the special character of the resources (plant, animal, marine, microbial resources) question the feasibility of such an instrument in the near future; and if such a treaty is signed, there is a fear that it will be a 'package deal' outcome and tormented with interpretation, implementation and enforcement problems. The task of the international community at the moment is not to draft another international treaty, but to detect methods to make functional the provisions of an already existing international treaty. In addition, input from international relations theory suggests that the international environment is not yet ripe for an international binding instrument in access and benefit-sharing.

Since the CBD, many countries have made serious attempts to enact domestic access and benefit-sharing measures, facing a range of legal and technical problems and a lack of experience in handling bioprospecting undertakings. Indeed, countries rich in biological resources cannot allow this opportunity to be lost but should take advantage of their biological wealth as they take advantage of any other valuable exported resources, such as timber, oil, mineral and subterranean resources. Bioprospecting should be received as an investment in biodiversity conservation in order to prevent logging and area clearance for arable lands.

However, source countries should not overestimate the value of their biological resources and should moderate their expectations, because unlike timber, oil, mineral and other resources, the value and use of which is almost certain, the development of a product from bioprospecting undertaking is uncertain. A heavy interventionist policy discourages future investors in biological resources and in the end all stakeholders are losers. More specifically, the cases of the Philippines and the Andean Pact have been analyzed. These cases were selected because they were the first complete access and benefit-sharing regulatory frameworks at the national and regional level respectively.

First, the Philippines have the most experience in access and benefit-sharing legislation. However, the long and complex procedures of the first access and benefit-sharing framework soon called for a second, more flexible one. The Philippines regulation was especially touted for its PIC procedure and for the accompanying access and benefit-sharing indigenous peoples' policy. Indigenous rights over land and knowledge are respected and indigenous peoples' stakeholders' status in the bioprospecting ventures is recognized. At the same time, good legislation has little value, if it is not coupled with an effective enforcement mechanism.

Secondly, the Andean Pact Decision on access to genetic resources was presented because to date it is the only regional instrument on access and benefit-sharing in force. In fact, the regional approach to access and benefit-sharing is highly recommended because neighbouring countries sharing biological wealth coordinate their access regulation efforts and thus their position in the international arena. The regional approach has two main advantages: on the one hand, it protects from the situation where the user of genetic resources avoids the country that enacted measures and pursues its activity related to the same resources in the neighboring country, where access and benefit-sharing regulation is absent. On the other hand, the users do not have to familiarize with many different regulations, since the main lines and principles are given at the regional level. Importantly, a regional approach in contrast to the international approach is easier to achieve because of the often similar mentality, cultures, and values in the region.

For this reason, a framework access and benefit-sharing regulation at the ASEAN level would further support the Philippines' efforts. Unfortunately, the Andean Pact framework has many flaws, and an amendment is indispensable to correct the inexistent PIC process, strengthen protection of indigenous rights and clarify its currently broad and unclear benefit-sharing provisions. Although the Decision was initially envisaged to be directly applicable, national action is necessary and the Andean Members have not responded in a coherent manner: the result is a multi-level and complex test of compliance for the users that deters bioprospecting undertakings in the area.

In the end, two issues have attracted the most attention in this study as critical for 'fair and equitable benefit-sharing': property rights and indigenous peoples, the former related to benefit-sharing with respect to land, genetic material, and traditional knowledge.

First, ill-defined land rights impede stakeholders' identification and render bioprospecting a very hard venture. Well-structured and defined land titles are promoted by academics and development institutions as a prerequisite for development. Finally, it is easier and more efficient to recognize, understand and try to integrate in bioprospecting undertakings the customary land and property systems developed from local and indigenous populations, than attempting to disqualify them and transplant Western models.

Secondly, with relation to the property nature of genetic material, the physical unity of the matter is important for a better investment environment. The separation of the matter in biological resource and genetic material, the state always being the owner of the latter, creates multiple parties which can deny access to outsiders, but none of them control enough property to work efficiently; contracts get complicated and investors discouraged. When genetic resources are extracted from protected areas, the state will be the main stakeholder as manager of these areas; but when the owner of the biological resources is an individual, a community or other entity, the State should supervise the whole process and protects the rights of its citizens in the arrangements.

With regard to traditional knowledge and intellectual property, there are efforts going on parallel to those for the construction of an international access and benefit-sharing hard law instrument, to amend the international patent standards, by including a requirement for disclosure of origin, source or legal provenance, benefit-sharing and PIC. It is believed that such an amendment would be costly and ineffective, causing confusion and bureaucracy. A bottom-up domestic approach is at the moment a preferable and more effective solution. Moreover, a compensatory mechanism is necessary for the contribution of the traditional knowledge holders in the medicinal R&D. It is difficult to imagine how a *sui generis* intellectual property tool can accommodate all forms of traditional knowledge because there are many conceptual problems to be answered which are related to the starting point for the protection duration and the fixation of the protection subject. In fact, it has been suggested here that broadening the scope of geographical indications and especially that of the *appellations d'origine* offers a good ground for offensive protection of local knowledge; while taking advantage of the remedies provided by current patent law and compiling databases and registers to assist prior art research guarantees defensive protection of traditional knowledge. The combination of these mechanisms forms a good conceptual basis for the compensation of the traditional knowledge-holders. Respect for traditional knowledge and practices should be supported and thus promoted by domestic policy: traditional knowledge is mostly held by indigenous populations and the lack of respect for indigenous rights at the domestic level fails the international respect of such rights. The compilation of databases and registers at a domestic level promotes the recognition of such rights and facilitates the compensation mechanism from the use of this knowledge.

Moreover, given that the majority of countries wealthy in biological diversity host indigenous communities that form part of the eco-system and manage these resources, indigenous peoples have rights and are entitled to further grant rights over the resources to be prospected. To meet the 'fair and equitable' criterion, access and benefit-sharing legislation is expected to recognize the stakeholder status of the indigenous peoples and facilitate their participation in the negotiation of the arrangement, guarantee their PIC and the return of some benefits. The rapidly developing international human rights body during the last decade further argues for the need to recognize the rights of these people, who in many countries for political reasons often remain ignored and are considered second-class citizens. It is hoped that the potent rise of the indigenous movement and the change in corporate culture will positively impact upon national legislations as well.

The focus on non-monetary benefits is critical for global sustainability concerns because developing countries receive training, technology and capacity building to add value and exploit their resources. At the same time, the need for monetary benefits remains important for the immediate financing of health-, education- and environment-related projects, and in this regard, trust fund mechanisms have proven rather successful in distributing benefits to and within local societies by investing on such projects.

Moving from source to the user countries, the scene is completely different. International law can be enforced only upon States, and here lies one of its main deficiencies. Under the present international economic order, transnational corporations are the main players in most aspects of trade, investment, and even politics by managing the majority of economic resources and they are not subject to a comprehensive international regulatory system. More specifically, it is easier to define how the provider State is expected to perform its obligations under the CBD. For the provider countries we know that for the sake of development and international justice, they have to design and systemize property rights and have to align their indigenous peoples' policy to the international human rights standards. However, for States with users, it is tricky to conceive of how they can enforce international treaties to which they abide upon a private sector that is seldom restricted within the borders of one state.

Hence, private companies' voluntary compliance with international treaties should be welcome, encouraged, and promoted. Private practices on access and benefit-sharing, and also indigenous peoples, are a very promising evolution in the international business environment and reflect the participation and reaction of these important stakeholders of the global economic integration in tackling international legal problems. However, voluntary codes of compliance adopted by the private sector do not amount to recognizing these private entities as subjects of international law. By setting common standards applicable in all states where the private company runs business activities, such voluntary behavior contributes to the harmonization of domestic legislation and fosters the evolution of private business law worldwide, but this is not sufficient to confer international legal personality to private entities.

States that host users and are thus entrusted with the regulation of these private sector stakeholders' activities should supply information and guidance, should assist with the implementation of ABS provisions, and in particular they should foster compliance with PIC and MAT. Countries with users under their jurisdiction should target public outreach, raising awareness, and dissemination of information. Certainly, raising awareness might appear a very poor instrument, if there are no administrative and judicial remedies available in the case of non-compliance with PIC and MATs, but it is argued here that the most effective way to enforce their obligations is by the better organization of research and technical and human resources support of their patent offices, in order to ensure against misappropriation of knowledge.

Access to genetic resources and benefit-sharing is a very complex issue and there is at the moment no single policy or regulatory answer to solve the various controversial aspects of the debate. Understanding the variable commercial demand for genetic resources, internally and externally;

industry needs; and the R&D partnerships requires better policies and regulatory frameworks in access and benefit-sharing, while case-by-case analysis is much needed to complement and inform policy making and legislative efforts. The CBD's strength might lie in leaving rules general, thus enabling different policy options and legal alternatives to be applied to each particular bioprospecting project.